DARWINISM

IN THE

ENGLISH NOVEL

1860–1910

DARWINISM IN THE ENGLISH NOVEL

1860-1910

The Impact of Evolution on Victorian Fiction

LEO J. HENKIN, PH. D.

NEW YORK
RUSSELL & RUSSELL · INC
1963

REISSUED, 1963, BY RUSSELL & RUSSELL, INC.
BY ARRANGEMENT WITH THE ESTATE OF LEO J. HENKIN
L. C. CATALOG CARD NO: 63—9505

PRINTED IN THE UNITED STATES OF AMERICA

TO MY WIFE

Submitted in partial fulfillment
of an eternal debt

ACKNOWLEDGMENT

I am glad of this opportunity to acknowledge my indebtedness to Henry Fairfield Osborn's *From the Greeks to Darwin,* Archibald Geikie's *The Founders of Geology,* and Robert H. Murray's *Science and Scientists in the Nineteenth Century* for the survey of evolutionary thought which constitutes the first and third chapters of this book; to Andrew D. White's *A History of the Warfare of Science with Theology in Christendom* for the facts of the controversy between evolution and religion; to *The Outline of Science* edited by J. Arthur Thomson, for the information about pre-historic man; and to James Osler Bailey's dissertation, *Scientific Fiction in English* for material on the scientific romance.

Above all I wish to express my gratitude to Professor Bruce McCullough of New York University for the guidance and encouragement he gave me from the beginning of this work to the end.

LEO J. HENKIN.

CONTENTS

INTRODUCTION

The purpose of this work is to treat the impact of the theory of evolution and the theory of natural selection upon the English novel. Although the date of the *Origin of Species,* 1859, establishes a starting point for such an inquiry, the evolutionary theory under the name of the "development hypothesis" began to press for recognition even before Darwin's reasonable explanation of how it worked gave the doctrine universal acceptance. Consequently, even before 1859, the novel touched on the theory, usually on the basis of geological and astronomical evidence. Passing consideration will be given to representative instances of such fictional treatment.

The major part of this study, however, will cover the fifty-year period following the publication of Darwin's epochal work: 1860-1910. Not until at least ten years after *The Origin of Species,* when the theory, weather-beaten but scathless, was first emerging from the storm of obloquy it had stirred up, did the novel begin to make use of Darwin's scientific doctrine. By 1910 the great controversy which had raged around the name of Darwin had long subsided. Evolution, then, had invaded every branch of science, ethics, philosophy, and sociology. As such it had entered so generally into the warp and woof of modern thought as to be indistinguishable as an independent factor.

It has been found impossible to admit into consideration the many American novels, and foreign novels and translations, which, contemporary with the English, reflect the current interest in evolution. This non-English fiction, however, duplicates for the most part the character and trend of the British product.

In general, the production of novels dealing with evolution as a theme or as a subject for exposition, was fertile. A number of causes contributed to this productivity. For one, the interest of Victorians in religion prepared the way for the success of theological and controversial fiction turning on the clash of Scripture and Science. The spiritual crisis which beset the Victorian Age today is but the shadow of a shadow; but for the contemporaries of Disraeli and Gladstone, it was a problem dramatically real, at once close and universal. It was a question of God or no God. Evolution in science meant revolution in religion. And the literature which chose to reflect to these earnest Victorians spiritual doubts by which untold thousands were beset, and a spiritual conflict in which blast and counter-blast sounded within earshot of all, was bound to appeal with all the force of the intimately personal and intimately familiar.

On another plane, the Victorian passion for science and practical information guaranteed a constant demand for popular literature in the form of the evolutionary novel. "To the world of the 1880's the story of life, of the origin and branching out of species, of the making of continents, was still the most inspiring of the new romances."[1] The Victorian generations that had formed the Mechanics Institutes, the Mutual Improvement Societies, the Athenaeums, and Philosophical Institutions were able to stomach meaty dishes; but the tasty stew of mixed science and romance purveyed by the novel proved especially palatable. Through the age the novel continued to be the chief medium of cultural communication, serving the end which the drama rendered in the Elizabethan Age and which the radio and moving picture are equipped to serve in the present.

These two phases of the Victorian interest in evolution—its effect on religion, and its contribution to the romance of science—constitute the chief developments in evolution's relation to Victorian fiction. Consideration of these developments will make up the larger body of this dissertation.

There are elements of perennial interest in the literature inspired by scientific imagination. Such literature may endure for a long time, sharing the regard paid to works endowed with timeless artistic imagination. But the novel that depends for its appeal on controversial issues dies as soon as these have begun to age. The religio-scientific question which fired the passions of all it touched in those soul-searching Victorian days is now cold as a fossil; and the novel which recorded those fears of faith in the presence of a godless science is now, with hardly an exception, of historical interest merely.

CHAPTER ONE

Survey of Evolutionary Thought to 1859

In November 1859 appeared Darwin's *Origin of Species.* The volumes marked an epoch in the development of modern thought. Since then a new race has come into the world. Everyone born since the year 1859 has breathed in the ideas and opinions which make up the philosophy of evolution.

When in the revival of learning the doctrine of development and gradual growth reappeared—for the idea in speculative form is as old as the Greek philosophers—it was as a philosophy that the concept was carried forward. Scientists had not yet collected sufficient data to treat the hypothesis of evolution as within the realm of fact. Consequently, it was as a philosophic idea of progress that the doctrine was passed on from the Renaissance humanists to the eighteenth century rationalists. Descartes, Bacon, Leibnitz, and Kant set forth evolutionary views, as philosophers approaching the subject from the metaphysical rather than from the scientific side, abstractly rather than empirically. But with the progress of experimental science in the eighteenth and nineteenth centuries evolution became the sphere of interest of astronomers, geologists, and biologists.

The sciences have developed in an order the reverse of what might have been expected. What was more remote from ourselves was first brought under the domain of law, and then gradually what was nearer: first the heavens, next the earth, then animal and vegetable life, then the human body, and last of all (as yet very imperfectly) the human mind. Consequently the doctrine of evolution dawned first in astronomy, then lighted up geology, and finally cast its full beams on biology.

Just as had the Copernican theory which revolutionized the traditional conception of the place of the earth in the universe, so the doctrine of evolution came face to face with theology. In the orthodox view, the world had been created in six days, and had contained from that time onwards all the heavenly bodies that it now contains, and all kinds of animals and plants, as well as some others that had perished in the Deluge. As a result of Adam's sin animals took to preying on each other, thistles and thorns grew up, and the very ground was cursed so that it no longer yielded sustenance to man except as the result of painful labor. This, it must be understood, was held to be literal historical matter of fact, either actually related in the Bible, or deducible from what was related. Further, a Protestant divine of the seventeenth century, Archbishop Ussher, had fixed for Protestant Christendom the date 4004 B. C. as the year of Creation.

Within this narrow framework science was expected to confine itself, and those who thought 6000 years too short a time for the existence of the visible universe were held up to obloquy.

Isaac Newton's work—the Copernican system having been accepted—did nothing to shake religious orthodoxy. Instead, from Newton the eighteenth century acquired its peculiar brand of piety, in which God appeared essentially as First Cause or Law-giver, who first created the world and then made rules which determined all further events without any need of His special intervention. The orthodox allowed exceptions—there were the miracles connected with religion. But for the deists everything, without exception, was regulated by natural law. Both views are to be found in Pope's *Essay on Man:*

> The first almighty Cause acts not by partial,
> but by gen'ral laws;
> The exceptions few.

The Reign of Law, as conceived in the time of Queen Anne, is associated with political stability and the belief that the era of revolutions is past. When men again began to desire change, their conception of the workings of natural law became less static.

In astronomy the belief that the heavenly bodies were unchanging gave place to the theory of their gradual development. As early as 1755, in a book called *General Natural History and Theory of the Heavens,* Immanuel Kant, then a young man of thirty-one, put forward the theory of a gradual transition in the universe from chaos to organization, beginning at the centre of gravity of the universe and slowly spreading outwards from this point towards the remotest regions—a process involving space and requiring infinite time. Pierre Laplace's famous nebular hypothesis was first published in 1796, in his *Exposition du Systeme du Monde.* Here he held that what is now the system of the sun and planets was originally a single diffuse nebula; that gradually it contracted, and in consequence rotated faster; that centrifugal force caused lumps to fly off, which became planets; and that the same process, repeated, gave rise to the satellites of the planets.

In geology belief in a former period of rapid and catastrophic changes was succeeded, as the science advanced, by a belief that change had always been very slow. At first it was thought that the whole history of the earth had to be compressed into about six thousand years. But the earth's surface and its strata bear witness to considerable changes. Since few of these changes have occurred during the last three thousand years or so of which history has kept record, earlier geologists, to accord with orthodox views, found it necessary to assume in the first three thousand years many vast and sudden catastrophies, whether natural or supernatural, enormously surpassing in their devastating power any convulsions of nature which have occurred in historical times. The stratifications of the rocks, they suggested, were the records of great cataclysms when the existing scheme of things was wiped out—a new page turned over. Writhing with seismic colic, the earth convulsed the mountains and belched the ocean upon the plains. This was the school of the Catastrophists or Convulsionists.

The eighteenth century was much occupied by a controversy between two Catastrophic schools: the Neptunists, who attributed almost everything to water, and the Vulcanists, who equally over-emphasized volcanoes and earthquakes. The former sect, who were perpetually collecting evidences of the Deluge, laid great stress on the marine fossils found upon the tops of the Alps. They were the more orthodox, and therefore the enemies of orthodoxy tried to deny that fossils were genuine remains of animals. Many held, indeed, that fossils were mere whimsies of the creator, stone patterns precisely imitating animal forms as frost on the panes fortuitously resembles moss or fern. Voltaire as protagonist of the unorthodox was especially sceptical. With the air of pricking another bubble he asked why it was necessary for some savants to sink the Alps under the sea in order to place a seashell on a peak. One had only to suppose that pilgrims on their way to Rome had carried shellfish in their picnic baskets to the mountains, and after feasting tossed the shells away. In this instance dogmatic free thought showed itself even more unscientific than orthodoxy.

It became realized in time that the shells, bones, and impressions of leaves found in stratified rocks could not be explained away as relics of animals drowned in Noah's Deluge or as "practice pieces" of creation, but were actual remains of life that had existed on the earth at the time the rocks were formed. About the year 1800 William Smith, so called "father of geology," made this crucial discovery. He found that when he examined cliffs here and there over wide stretches of territory, the rocky layers were almost always in the same order. Beneath the upper chalk layers was a layer of clay, next lower came strata of green sandstone, and below these were several layers of oolitic limestone. Wherever the crust of the earth was exposed these layers were of the same relative thickness and in the same order. Not only this, but the same kind of fossils were found in corresponding strata over tremendous areas. What more natural conclusion than that such a layer as the old red sandstone, wherever found, belongs to the same age, and the fossils imbedded in it belong to the same period? However, it was still believed that the life represented in each formation was a separate creation wholly independent of what had gone before. Destruction of existing life by an assumed general catastrophe was followed, the theory ran, by creation of the slightly different organisms found in the next younger beds.

In other parts of the world we find most of the sequence of rocks and fossils noted by Smith but with several strata of newer rocks above it or with older strata beneath it. Not all the rocks that have been laid down, much less all fossils deposited in them, have been permanently preserved. Rocks are worn down and destroyed by rains, winds, waves, glaciers and other agents of erosion. Fortunately, however, when a given stratum is being destroyed in one region it is being preserved in another. On the whole, therefore, some part of nearly every chapter in the history of the earth is available for reading. Sometimes some important context is missing, but more often the significant pages have been preserved. Indeed, some of the chapters in the record of the rocks are nearly complete.

The first writer to set forth a modern scientific view in geology was James Hutton whose *Theory of the Earth* was first published in 1788, and in an enlarged form in 1795. At a time when the consequences of the French Revolution had created a British opinion haunted by dread of innovation, Hutton invited the onslaughts of the upholders of Genesis by attributing the vast changes which have occurred during geological time to those very causes which are now observed to be slowly altering coast-lines, or increasing or diminishing the height of mountains, and raising or lowering the ocean-bed. By this theory he started a definite trend toward a general lengthening of geologic time, traveling backward from 4004 B. C.

In his *Principles of Geology* published between 1830 and 1833, Sir Charles Lyell followed the doctrines of geological uniformity and continuity laid down by Buffon and Hutton before him, settling for all time the violent dispute which the theories of his predecessors had started. By Lyell's day the argument had already proceeded a considerable distance from 4004 B. C. But it still remained a moot point whether the major geological effects manifest in the earth's surface could not have been produced by great and sudden cataclysms in a relatively short time. Thus the whole issue reduced itself to a question of fact: whether the known conformations had been produced by cataclysm. It was Lyell's view that *no* cataclysm could or did produce any major geological formation. There is no evidence whatever, he held, that the dynamics of the earth have ever been different in any essential respect from what they are today. His work swept away theories of cataclysms and catastrophes. With a bold simplicity he underlined his thesis that the revolutions through which the surface of the globe has successively passed—whether by the elevation or submersion of continents, the formation of mountains and valleys, the hollowing out of water-courses and lake beds—were brought about gradually and slowly during incalculable lapses of time by such natural processes which we see going on around us in the present age. By his Uniformitarian theory Lyell revolutionized geology and assisted in paving the way for the removal of one difficulty attending the solution of the theory of the origin of species, namely, the vast period of time for the life history of the globe which that theory demands.

As a result of the investigations of Dr. Louis Agassiz, who propounded the theory in 1837, and of subsequent research, it soon was considered an established fact that at a period in the past roughly corresponding with the Quaternary period (Pleistocene epoch) the earth had been subjected to a long reign of ice and snow, in the northern and southern hemispheres alike. This is the Glacial Epoch theory. During this Ice Age the temperature in many parts of the world grew so cold and the snowfall so heavy that in mountain regions the permanent snow extended several thousand feet down below its present line, while the glaciers were prolonged for some hundreds of miles down their valleys. There is, indeed, abundant evidence that the glaciation, at least in the northern hemisphere, was not a single episode but comprised four and possibly five glacial stages, separated by intervals during which the ice disappeared from most of the area that it had covered, and the climate was much like that at present. During such interglacial stages, the surface from which the ice had melted was again subject to weathering and erosion, new soil was formed, and the land was reoccupied by plants and animals.

The further progress of geology is entangled with that of biology, owing to the multitude of extinct forms of life imbedded in the earth of which fossils preserve a record. And as the discovery of fossils proceeded it became more and more difficult to reconcile the existence of all these huge animals with theology and the Mosaic record.

Trouble arose too from the mere number and size of the species that came to be known with the progress of zoology. The numbers now known amount to millions, and if two of each of these kinds, "every creeping thing that creepeth upon the earth, and every fowl," were in the ark, it was felt that the ark must have been rather overcrowded. Theology's views on animal life were uncompromising. In those days the doctrine of the Special Creation of species was an idiom that a child acquired before he was ten years old; even a naturalist would no more suppose that one species or sub-kingdom gave rise to another than that Italy gave rise to Spain. Men believed that every existing animal, however numerous in kind, belonged to species represented in Noah's ark; that all those extinct, discoverable in fossil form, had been with few exceptions drowned in the Noachian Deluge.

Difficulties had begun early with the discovery of the new world. America was a long way from Mount Ararat, yet it contained many animals not to be found at intermediate places. How came these animals to have travelled so far, and to have left none of their kind on the way? With a million known species to account for, the traditional explanation of a supernatural creation of each of them became increasingly difficult to accept. One theory—called in to the aid of the divine fiat concept—held that many species had been spontaneously generated out of slime by the action of the sun. But a genuine resolution of these difficulties was to come only through the science of biology and the theory of evolution.

The doctrine of the gradual evolution of plants and animals by descent and variation came into biology largely through geology. There was the fact, as fossilized strata indisputably showed, that the simpler forms of life were the first to appear, and those with a more complicated structure arose at a later stage of the record. Further, the general similarity in structure between fish and reptile, bird and beast, naturally suggested a connection by way of descent or development.

Sixty years before the birth of Darwin, Georges Buffon, the great French naturalist, gathered together all that men knew at the time about the animal kingdom. Buffon could not be content with the mere collection and analysis of his material. He generalized about this amazing store of facts with marvelous foresight, and although the statement is unquestionably exaggerated, Samuel Butler held that Buffon had anticipated all that Charles Darwin presented to his generation in the nineteenth century.[2] Buffon put forth the view that species have altered, owing to changes in the environment in past ages of the world; and with a philosophic eye observing the unity of nature he hinted that there may have been a common ancestor of the ass and horse, and of the ape and man.

As much for these views as for others in the field of geology on the factors at work sculpturing the earth, the Sorbonne theological faculty in Paris condemned him on fourteen points. Buffon recanted and was obliged to publish his retraction in full in the next volume of the *Natural History* (1751):

> I declare (it reads) that I had no intention to contradict the text of scripture; that I believe most firmly all therein related about the creation, both as to order of time and matter of fact; I abandon everything in my book respecting the formation of the earth, and generally all that may be contrary to the narration of Moses.[3]

It is interesting to note that another great naturalist exactly contemporaneous with Buffon, Karl von Linnaeus, a botanist and the first great classifier of plant life, was a believer in the absolute fixity of species almost to his death. Species were in his mind the units of direct creation; each species bore the impression of the thought of the Creator in all its structure and function. But Linnaeus was compelled later in his life to give up, in some little measure, this idea of the Special Divine Creation and absolute fixity.

Across the channel, an English physician, seeking out a theory of disease which should have a scientific foundation hit upon conceptions concerning the factors or causes of evolution strangely like those afterward proclaimed independently by the more scientific Jean Lamarck. To students of heredity it is a fact of surpassing fascination that Dr. Erasmus Darwin, grandfather of the great naturalist, should prove to be a poet of evolution. He wrote the *Botanic Garden,* two volumes of verse published about 1788; the prose *Zoonomia,* published in 1794; and the *Temple of Nature,* published after his death in 1802.

Dr. Darwin was more than a mere follower of Empedocles and Lucretius. As a student of science he had read the writings of Hutton in geology and undoubtedly derived something from Buffon and Linnaeus. But where Charles Darwin was to catch and hold views, Erasmus Darwin, a philosophic poet principally, caught only glimpses. Having had occasion to note a certain resemblance in structure which obtained in all warm-blooded animals, he wondered if all organic life might not have been produced from one and the same kind of living filament. Further, arguing for the mutability of species, he made the first clear and definite statement of the theory of the transmission of acquired characters: "All animals undergo perpetual transformations which are in part produced by their own exertion in consequence of their desires and aversions; of their pleasures and pains; or of irritations and associations; and many of these acquired forms are transmitted to their posterity."[4] This deduction appeared in Darwin's prose *Zoonomia* (1794-6), fifteen years before Lamarck's *Philosophie Zoologique* (1809) offered an identical thesis.

Zoonomia, or the Laws of Organic Life, sets forth completely in a short sketch the principles of evolution, indicating their clear development in the elder Darwin's mind. Arguing from the evidences of artificial selection, of similarity of structure, of change, of adaptations to environment, and allowing for "a great length of time, since the earth began to exist perhaps millions of ages before the commencement of the history of mankind,"[5] Dr. Darwin propounds a theory of descent no less remarkable that it is more speculative than scientific. More, in another prose work, the *Phytologia* (1799), he describes the fierce struggle for existence in which animal destroys animal and plant destroys plant. "Such," he exclaims with a grim irony that reminds one of Schopenhauer, "is the condition of organic nature! whose first law might be expressed in the words 'Eat, or be eaten!' and which would seem to be one great slaughter-house, one universal scene of rapacity and injustice."[6] The philosopher-poet-physician saw in this bitter contest, like Thomas Malthus a few years later, checks to the naturally rapid increase in life; but he failed to take the next step that Charles Darwin was to take of envisaging in the struggle a sifting process for the improvement of species. But in one expressive passage the elder Darwin embodies in words all but identical with the famous formula of the survival of the fittest, the doctrine of sexual selection: "The final cause of this contest among the males seems to be, that the strongest and most active animal should propagate the species, which should thence become improved."[7]

Darwin's poetizing of these ideas brought upon him the scoffing of J. H. Frere and George Canning of the *Anti Jacobin*[8] and of S. T. Coleridge in the *Biographia Literaria;*[9] and, if we are to accept the statement of Samuel Butler, the classical example of the argument for design in the universe, Paley's *Natural Theology,* was "written throughout at the 'Zoonomia'."[10] Even Charles Darwin, who had read the work with admiration as a youth, disparaged it on second reading because of the large "proportion of speculation,"[11] prejudiced against it no doubt by its adumbrations of Lamarckism which he detested.

It is one thing to improvise hypotheses and theories out of the fulness of one's fancy, even when supported by a considerable knowledge of nature; and another to demonstrate them by an enormous number of facts and carry them to such a degree of probability as to satisfy those capable of judging. Dr. Darwin was a theorist and scarcely more; generations ahead of his contemporaries, he could not satisfy them with his physio-philosophical ideas. Without a scientific portfolio he could not claim a hearing for his theories even from the scientific world. When Lamarck set forth his version of development he did so in complete ignorance of Darwin's anticipation of his doctrine, and with complete faith in its originality. And since the French naturalist supported his theories by facts, science accorded Lamarck the distinction of labelling doctrines "Lamarckism" to which Erasmus Darwin had prior claim.

It was in 1809, the year of Charles Darwin's birth and exactly fifty years before the appearance of the *Origin of Species,* that the French naturalist Jean Lamarck declared after nine years of wrestling with the immense invertebrate world, that the species of the animal kingdom form a connected series, a graduated chain from the monad (one-celled organism) to man. But the year 1809 was full of sound and fury, and with eyes following the dust of Napoleon's columns, Europe had none for the *Philosophie Zoologique* and its startling concept of life.

Nevertheless, Lamarck's two volume work may be called the first systematic and logical presentation of the evolutionary theory in biology. Lamarck believed that the desires and efforts of an animal to adjust itself to changes in the conditions of life—such as climate, food, etc.—result in the production of an organ with a new function. In this way a new animal is produced. Thus, when mammals entered the strange new environment of the sea, they took on fish-like forms. This modification of form, in his opinion, is due to the well-known physiological law that all organs are strengthened by constant use, while they are weakened or even completely lost by disuse.

Disuse of an organ through generations, he reasoned, will result in its atrophy, perhaps its disappearance; constant development of some part of the body will result in the exaggeration of that organ until in the course of time such change is stamped upon the race. He remembered how the leaves of the ilex oaks by the Mediterranean were reduced in size, thickened and felted to cut down the loss of water in the fierce southern sunlight. The hooves of the mountain goat are dainty for climbing upon narrow ledges; the foot of the horse is adapted to the steppe. And—most classic example—the giraffe got his long neck by striving to reach the higher leaves on the tall palm trees. Charles Darwin's suggestion, that the antelopes with the long necks survived in conditions under which the short-necked ones perished, meets with more general approval.

Today Lamarckism is still a recognized school of evolutionary thought, though the transmission of acquired characteristics is not a theory generally accepted by modern zoologists. In its own day, despite the support of Geoffroy St. Hilaire, Lamarckism was crushed beneath the heavy weight of the authority of Cuvier, the great naturalist and comparative anatomist of that time. It fell almost stillborn. Its birth was premature, for the world was not yet prepared for it.

Scientists found chief fault with Lamarck's *transformisme* in the supposed efforts of species to adapt themselves by acts of repeated volition to altered conditions surrounding them. This factor seems to have taken in the mind of Lamarck the somewhat vague and transcendental form of aspiration, an upward striving of the animal toward higher conditions. Darwin scolded Lyell for referring to natural selection as a modification of Lamarck's doctrine of development and progression. Aside from Lamarck's holding that species descended from other species, Darwin could see nothing "in common between the

'Origin' and Lamarck"; and begged Lyell to desist from linking his views "with what I consider, after two deliberate readings, as a wretched book, and one from which (I well remember my surprise) I gained nothing."[12] Yet, for all this scorn, this very phase of Lamarck's hypothesis was selected by Samuel Butler, and in turn by George Bernard Shaw, for special defence and emphasis.

Sir Charles Lyell's *Principles of Geology* (1830-1833) contained an elaborate refutation of Lamarck's views concerning the progressive development of species. For Lyell was not an evolutionist at the time of the publication of his *Principles,* and he succeeded in maintaining a judicious reserve on the burning issue almost up to the time of his death in 1875. Nevertheless, despite his orthodoxy, in the field of biology as well as in geology, Lyell had set in motion the evolutionary impulse. Where the old school saw miracles and sudden creations, the new school began to see slow development and substantial continuity throughout enormous periods of similar activity.

This development coincided with Charles Darwin's *Lehrjahre.* Lyell's work had not yet been published in 1828 when Charles Darwin entered Christ's College, Cambridge. That college could then have claimed that on its bead-roll it had the name of John Milton, the writer of the epic of the Special Creation theory, which another of its sons, Charles Darwin, was to destroy. However, early in 1831, it was with unshaken belief in immutability and the creationist point of view that Charles Darwin left England on a long voyage of scientific exploration on H. M. S. *Beagle* (1831-1836). He took with him a copy of Lyell's *Geology.* The young naturalist studied it attentively. He thrilled at the thought of applying the principles of the great Scotsman to the geology of the countries he was about to visit.[13] Inevitably, too, there began to stir in his mind other thoughts intimately related to Lyell's argument. If such forces as are now at work were capable of raising the continental masses and carving them down to their present contours, given enough time, why should not other forces such as are now at work on the animal kingdom be capable over a similar span of time of producing the differences which distinguish the successive species? This was the question which Lyell unintentionally set for Darwin. Darwin was soon completely convinced of the deep significance of the transmutation of species, and set himself to seek out the laws of change.[14] In July, 1837 he opened his first notebook. The *Origin of Species* had struck root.

Between Lyell's *Principles* and Darwin's *Origin* the evolutionary landmarks are two: the work of Robert Chambers, and that of Herbert Spencer.

The celebrated *Vestiges of the Natural History of Creation,* published anonymously (it was attributed, by a wild surmise, to Prince Albert), but acknowledged later to have been written by Robert Chambers, preceded the *Origin of Species* by fifteen years. *The Vestiges* is now generally associated with an abortive theory of the origin of species eclipsed by the solid achievement of Charles Darwin. But its historical importance is independent of any particular biological theory. Chambers really founded the doctrine of evolution in Britain, and created the atmosphere which made free discussion of man's origin a possibility.

Darwin himself admitted that the comparative rapidity with which the educated public accepted, or at least tolerated his views was greatly due to the previous publication of the *Vestiges* in 1844 and in nine successive editions before 1854.[15] Chambers' work was less scientific than philosophic and its popularity derived principally from its appeal from the too technical court of science to the wider court of popular intelligence. It was the first, therefore, to make the English public acquainted with the idea of evolution as applied to the history of the earth and its forms.

The *Vestiges* traced the origin of the solar and stellar systems and expounded the nebular hypothesis. It sketched the earth's history according to the uniformitarian theory advanced by Lyell, then went on to attack the doctrine of special creation, and to uphold an evolutionary creed. Every change through which the conformation and inorganic constitution of our planet had passed in the course of ages was explained by natural causes. Chambers used the arguments now familiar but then generally unknown, from the survival of rudimentary structures; from the development of the embryo which in rough fashion recapitulates in its own growth the earlier stages of organic advance; from the facts of geographical distribution of organisms; from the unity of structural type among species. Life itself he was ready to regard as a development from the inorganic, brought about by natural causes, without any special divine interference. To account for the fact of progress he suggested the hypothesis that, at certain points in the evolutionary process, the embryo of some species would suddenly take a structural leap forward, and render possible the emergence of a more highly developed organism. This theory orthodox Darwinism was to reject, on the ground that *Natura nihil facit per saltum.* It is important to note, however, that the occurrence of "mutations" and "discontinuous variations" has today been established.

The *Vestiges* was violently attacked both by men of science and by theologians. Fixity of species was still the orthodox scientific creed, and was firmly upheld by such authorities as Professors Adam Sedgwick and Louis Agassiz. Theologians objected to the book on two grounds that were soon to become common: first, that it contradicted *Genesis* which was held to teach authoritatively the doctrine of Special Creation by divine *fiat;* secondly, that evolution meant materialism, the suggestion of an animal ancestry for man being regarded as a degrading belief.

Chambers had grasped the various lines of reasoning by which the theory of evolution has since been established and has won universal acceptance. He saw what men like Agassiz and Sedgwick were blind to and denied to the day of their deaths—the origin of organic species by modification of pre-existing types.

At Down, Charles Darwin, home from the cruise of the "Beagle," had been inspired to formulate a theory on the origin of species by his reading in 1838 of Malthus on *Population.* In January, 1844 he wrote to his friend Sir Joseph Hooker: "At last gleams of light have come, and I am almost convinced (quite contrary to the opinion I started with) that species are not (it is like confessing a murder) immutable."[16] Yet having carefully read the *Vestiges* he could, strangely enough, write: "The idea of a fish passing into a reptile, monstrous." And Darwin was not exactly flattered that some had attributed the book to him.[17]

Herbert Spencer, Darwin's contemporary and often popularly regarded as one of Darwin's disciples, was in point of fact an avowed evolutionist long before the publication of Darwin's first hint on the subject. Seven years before the *Origin of Species,* in 1852, he published an essay entitled *The Development Hypothesis,* in which the "theory of evolution" is fully and closely set forth both in concept and in name; and in 1858, Spencer issued a program for a series of works dealing with astronomic evolution, geologic evolution, the evolution of life in general, and evolution in individual organisms.

The cosmic evolutionism of Spencer was clearly foreshadowed by the progress of evolutionary thought before his time. Kant and Laplace had worked out the development of suns and earths from white-hot star clouds; Lyell and the geologists had worked out the evolution of the earth's surface to its present condition; Lamarck had worked out the descent of plants and animals from a common ancestor by gradual modification. But all these thinkers had addressed themselves to one set of phenomena alone; had regarded the process which they pointed out in isolation only. It remained for Herbert Spencer with a philosophic view of evolution as a cosmical process—one and continuous, from nebula to man, from star to soul, from atom to society—to synthesize and unify these scattered evolutionary concepts. In his hands the principle of evolution was made the key to the understanding at once of the inorganic and organic world, of the human mind, and of the intricacies of social life.

And then came *The Origin of Species*.

CHAPTER TWO

Evolutionary Thought in the Novel Before 1859

When Darwin had published his epochal books *The Origin of Species* and *The Descent of Man,* and sharp controversy was being waged regarding his views on the pedigree of man, a witty Scottish lord proclaimed in a set of verses a popular notion which had long been current in the northern kingdom, viz., that Darwin had had a forerunner in the person of a highly eccentric Scotch senator of the eighteenth century—James Burnet, Lord Monboddo. Lord Neaves' stanzas *To the Memory of Monboddo* pay posthumous tribute to the Scotch judge who had anticipated the theory of man's fundamental relationship with the higher apes:

> His views when forth at first they came
> Appeared a little odd, O!
> But now we've notions much the same
> We're back to old Monboddo.
>
> The rise of Man he loved to trace
> Up to the very pod, O!
> And in Baboons our parent race
> Was found by old Monboddo.
>
> . . . .
>
> Though Darwin now proclaims the law
> And spreads it far abroad, O!
> The man that first the secret saw
> Was honest old Monboddo.[1]

Thomas Love Peacock studied Lord Monboddo's *Ancient Metaphysics,* which remained one of his favorite books to the end of his life, and the same author's *Origin and Progress of Language,* a ponderous work in six volumes (1773-92).[2] The novelist's particular interest in Monboddo's treatment of the ape's relation to man is embodied in his documented synthetic novel, *Melincourt* (1817).[3]

Lord Monboddo was firmly convinced of the humanity of the orang-outang (man-of-the-woods), a species of ape. Orangs are, he said, further advanced than many savages found in other parts of the world, and even of Europe, inasmuch as they invariably walk upright—not like many of the savages, upon all fours. He collected evidence to show that the orang was capable of using a stick for defense as well as for attack, could learn the business of a common sailor, could be taught to play on the flute, was capable of great attachment to particular persons, and was orderly and dignified in behavior, loathing the habits of restlessness and destruction characteristic of the monkey. "If," said Monboddo, "such an Animal is not a man, I should desire to know in what the essence of a Man consists, and what is it that distinguishes a Natural Man from the Man of Art?"[4]

Further, in his *Origin and Progress of Language,* the Scotsman went beyond his *a priori* assertion that the orang-outang is of the human species. There he held that man was once an animal, without speech, or reason, or affection; that originally man possessed a tail which he ultimately lost by the constant posture of sitting; that men with tails did still exist; that men and orang-outangs are of the same species, the latter having only the capacity of speech while men have learned to make actual use of it.

Daring indeed was it in that day, especially in Scotland, to broach a theory which implicitly denied the Special Creation of man. But the sting of Monboddo's book was in its tail. He was mourned over by the pious, laughed at by the wits. Dr. Samuel Johnson found him always a source of amusement. Most men, observed Johnson, endeavor to hide their tails; but Lord Monboddo was as vain of his as a squirrel;[5] and Sir Adolphus Oughton succeeded in restoring Johnson to halcyonic temper during a heated discussion by calling Lord Monboddo a "Judge *a pos-*

teriori."[6] Had Monboddo lived till our day when comparative embryology is an established branch of biology, he would have had his chance to laugh at the wags. He would have found unsuspected support for his anthropological speculations in the fact that although any outward and visible sign of the human tail has disappeared, the human embryo supplies proofs that man's remote ancestors possessed one. The remnants of his caudal appendage, a few small vertebrae beneath the skin, are among the seventy and odd vestigial structures which make of man a walking museum of relics, enigmatical except in the light of the past.

Monboddo was by no means the only one in pre-Darwinian times to see kinship between the species man and the simian. Thus Washington Irving, in his *History of New York* (1809), mentions "the startling conjecture of Buffon, Helvetius, and Darwin, so highly honorable to mankind, that the whole human species is accidentally descended from a remarkable family of monkeys!"[7] Thomas Love Peacock's knowledge of the hypothesis was derived principally from Monboddo's lucubrations, but Peacock made use as well of the works on the human species by Linnaeus (*Systema Naturae,* 1735) and by Buffon (*Histoire Naturelle,* 1749-1788). Linnaeus seems to have regarded the orang-outang as a species of man, classifying him as *Homo nocturnus* and *Homo silvestris.* Buffon, in turn, in his account of "Les Orang-outangs ou le Pongo et le Jocko" in the fourteenth volume of his great work (1766), though placing the orang among the monkeys, accredits him with almost all the characteristics of the human species.[8] Drawing from Linnaeus, Buffon, and most frequently of all from Monboddo, Peacock composed an Aristophanic comedy centering around an ape with human attributes, justifying each trait by citing chapter and verse in the works of these famous scientists.

In an earlier novel, *Headlong Hall* (1816),[9] written around the question whether the human race tends toward moral perfection, Peacock has one of the characters, the deteriorationist Mr. Escot, exclaim: "Give me the wild man of the woods; the original, unthinking, unscientific, unlogical savage; in him there is at least some good."[10] This passage leads naturally on to *Melincourt*. The grand central feature of *Melincourt* is precisely such a savage, an orang-outang caught young in the forests of his native Africa who has been introduced to the best English society, of which as Sir Oran Haut-ton he becomes a distinguished ornament.

Mr. Forrester, patron of the gentle simian, contends that the orang really is a member of the human family whom he wishes to see established among his brothers who have the gift of speech. To this end, by means of a sum of money delivered to the right person, Mr. Forrester has purchased him a baronetcy and proposes to get him returned to Parliament for the rotten borough of One-vote. The quadrumanous baronet lends himself willingly to be educated and domesticated. Of honest instincts, considerable sobriety, and a keen talent for music, Sir Oran needs only the gift of speech to justify Forrester's exalting him to a kingship of the world. Sir Oran has the manners of the best company. He is of profoundly grateful disposition, and shows himself as capable of love as of friendship; he is heroically chivalrous when occasions present themselves, engaging in adventures worthy of a paladin of Charlemagne. Thanks to his extraordinary personal strength he rescues Mr. Forrester's beloved when she is placed in jeopardy by a mountain torrent: Sir Oran bridges the raging stream with a riven pine tree which he handles with the ease of a Paul Bunyan. His extreme reserve passes with the many as indicating a powerful but cautious thinker. Only occasionally does the mild-mannered ape break bounds, as when, after behaving through dinner with extreme decorum, suddenly, after slightly exceeding in madeira, he takes a flying leap out at the open window and "goes dancing along the woods like a harlequin."

The ingenuous imitativeness of the man-of-the-woods raises him to within an inch or two of the human platform, while the simple courtesy of his nature raises him high above it. With a pen dipped in Swift's caustic, somewhat diluted, Peacock, half in jest, half in earnest, takes the opportunity to point out many unflattering similarities and still more unflattering contrasts between Sir Oran and his associates; especially opposing the character of the natural man to the meanness of Lord Anophel Achtha, the son of a hundred earls, and to his creature, the Reverend Dr. Grovelgrub.

But the point of the satire lies in the application of the footnotes. There is not a quality or an action attributed to Sir Oran that is not based upon grave extracts from writings by Linnaeus, Buffon, and Monboddo. Collecting and welding their statements, Peacock uncovers a scientifically established, ideally perfect man; a Noble Savage who, though (or rather, because) unacquainted with the degenerate arts of civilization, has all the instincts and puts into practice all the precepts of true nobility. The author quotes Linnaeus on the orang-outang: "He has an upright gait and hissing speech; he believes that the world was created for him, and that he will one day rule over it again."[11] With dry humor Peacock presents to contemporary civilization its own scientific discovery, and seems bent on persuading that the unobtrusive, well mannered gentleman with the dark complexion and the bushy whiskers was really progressing toward perfectibility from an intellectual order of apes.

This he did, it may be, with his tongue in his cheek. Though just what the satirist's object was in creating the role of an educated orang-outang is not quite clear, it is reasonable to suppose that Peacock endowed Sir Oran with so much goodness, albeit scientifically, merely to reduce to an absurdity Monboddo's doctrine that the ape is human, unjustly degraded by the zoologists to rank below man. But while Peacock may have been laughing only at Lord Monboddo's theories, we cannot be too sure that he was not also laughing a little at the people whom those theories made hot with anger. Certainly it were best to regard *Melincourt* merely as a bit of whimsical fooling, lest we fall into the same pit with Lord Byron who had the conceit to fancy the tale was a skit[12] at his bear.*

A year later, in 1818, Peacock published another *jeu d'esprit, Nightmare Abbey.*[13] In this, while setting up caricatures of Shelley, Byron, and Coleridge in the stocks, the satirist finds occasion to pillory an absurd and credulous scientist who "maintains the origin of all things from water" and has spent his whole life in searching for specimens half-human and half-fish, "the orang-outangs of the sea."[14] This gentleman, the ichthyologist Mr. Asterias, is persuaded that these sea-men differ in no respect from land-men except that they are stupid and covered with scales. On the watch for such a triton, Mr. Asterias and his son Aquarius capture in the meshes of their net the pessimistic Mr. Toobad, a guest at Nightmare Abbey.

Darwin himself during this time was revolving slowly in his mind the observations and ideas which had come

---

* There is no doubting the point of the jest intended by Peacock in his novel *Gryll Grange* (1861), written shortly after *The Origin of Species* and other works on the antiquity of man. In the novel one of the characters presents a lump of chipped flint found among the bones of mammoths, and remarks that scientists consider such weapons of human manufacture evidence of a greater antiquity of the human race. The Reverend Dr. Opimian, Peacock's mouthpiece in the novel, finds this enormously amusing:

"This beats the Elephant in the Moon [Butler's poem] which turned out to be a mouse in a telescope. But I can help them to an explanation of what became of these primeval men-of-arms. They were an ethereal race, and evaporated."

Thomas L. Peacock, *Gryll Grange* (London, 1896), pp. 163-4.

to him in the course of his voyage as naturalist on the *Beagle.* But late in 1846 his attention had been turned aside to a work on "Cirripedia," an order of crustaceans, and for eight years work on this subject consumed the greater part of his time. Ultimately this microscopic task was finished and Darwin published two thick volumes describing all the known living species, and two thin quartos on the extinct species.[15]

Both the *Beagle* voyage and *Cirripedia* volumes earned Darwin obscure niches in works of fiction. In *Wives and Daughters* (1866) Mrs. Gaskell, a friend of the Darwins, is said to have represented in the hero of that work, Roger Hamley, something of the character and experiences of Charles Darwin, somewhat romanticized.[16] Roger is a scientist, "full of his natural history and comparative anatomy," a plodder who never realizes his own real worth. Thwarted in love he goes on an expedition which recalls that of Darwin as recorded in *The Voyage of the Beagle.*

The publication in 1844, fifteen years before the appearance of Darwin's *Origin of Species,* of Robert Chambers' *Vestiges of the Natural History of Creation,* it has been remarked, stirred the legions of militant orthodox Christianity into violent antagonism. In this daring speculation, Chambers had placed the issue squarely between two rival theories, the theory of creation by law, and the theory of creation by a number of miraculous interferences with the course of nature; and had cast his ballot for creation by uniform law. He had tried to show that the world constituted one vast self-acting machine by which everything from stellar systems to animalculae was turned out as it was required with universal and undeviating regularity. Despite his argument that the production of such a machine did more credit to the divine wisdom and power than a state of things necessitating a perpetual readjustment of means to ends, contemporary theologians regarded the devout author of the *Vestiges* as a dangerous enemy.

Benjamin Disraeli, M.P., by then an established novelist, read Chambers' work and was "enchanted with it";[17] but he ranged himself with the theologians just as soon as he realized in this abortive manifestation of the evolutionary thesis an attempt to dethrone the Spiritual. A Conservative in religion and politics, by temperament and persuasion both, the future Earl of Beaconsfield held firm belief in the necessity of fixed dogma, for the peace of mind no less than for the peace of State. More than a passing show of spleen therefore was his deliberate sidestep in the novel *Tancred* (1847),[18] published three years after the *Vestiges,* to dig his heel into Chambers' evolutionary hypothesis.

Tancred traces the youth of a high-minded idealist who desires to effect the regeneration of the West by a restoration of faith. Moved by discontent, on attaining his majority he astonishes his parents by foregoing a career in Parliament, and announcing his intention of making a pilgrimage to the Holy Sepulchre at Jerusalem. Every effort to dissuade him is made by his father, by a bishop, by a man of the world, but in vain. Finally he is introduced to the allurements of London society where a chance meeting with the beautiful young Constance Rawleigh inclines him to postpone his journey.

Lady Constance fascinates Tancred and becomes in his thoughts a serious rival to Jerusalem. Nursing the religious dreams he longs to share with her, he finds her to his dismay reading *The Revelations of Chaos,* a new book which she insists "explains everything." The spell is broken.

There is no missing Disraeli's biting tone when Lady Constance in a garrulous overflow summarizes for Tancred the arguments of Robert Chambers' *Vestiges of the Natural History of Creation* masquerading as *The Revelations.*

"You must read the 'Revelations'." Constance insists; "it is all explained. But what is most interesting, is the way in which man has been developed. You know, all is development. The principle is perpetually going on. First there was nothing, then there was something; then I forget the next, I think there were shells, then fishes; then we came, let me see, did we come next? Never mind that; we came at last. And the next change there will be something very superior to us, something with wings. Ah! that's it: we were fishes, and I believe we shall be crows. . . ."

"I do not believe I ever was a fish," said Tancred.

"Oh, but it is all proved. . . . This is development. We had fins—we may have wings."

"I was a fish, and I shall be a crow," says Tancred to himself.

Sadly he walks away. "What a spiritual mistress!" he exclaims, withal relieved at his escape. "And yesterday, for a moment, I almost dreamed of kneeling with her at the Holy Sepulchre."[19]

For all of Disraeli's lampooning, Chambers anonymously continued to publish edition after edition of his pseudo-scientific treatise, thus keeping the question of Law versus Miracle alive until it was ripe for Darwin's solution.

In 1849 appeared Robert Hunt's *Panthea* or *The Spirit of Nature.*[20] Hunt had already gained fame as the author of *The Poetry of Science,* a popular treatise on chemistry and related sciences similar in nature and object to the *Kosmos* of Humboldt. But *Panthea* is of very different character. It is a poetical romance exhibiting the "progress of a young and ardent mind, captivated by the beautiful in Nature, and allured by the wonders of Science."

Laon Elphage, a mystic philosopher, acquires a commanding influence over the mind of Lord Julian Altamont. The brilliant young nobleman could not be restrained from associating with Laon and his equally mystical-minded and accomplished daughter Aeltgiva, an enthusiastic votary of Panthea, the spirit of nature. Under their unorthodox guidance, the young heir of Altamont is ushered into the presence of Panthea who unfolds to him all the mystery of creation. She exhibits the various phases of the earth's history from formless chaos to the appearance of man upon the stage of life, then up the various steps of human progress. After a severe illness, however, Julian awakes from such an illusory state of existence, shakes off the incubus, and becomes like Faust a worker for the people.

Julian did not come altogether a novice into the hands of his rosicrucian preceptors. His studies in the natural sciences had taught him many things that served as a key to Panthea's revelations. His studies in geology, strictly according to Lyell, had revealed to him the gradual development of the earth and the short span man has inhabited it, "the existence of his race being but as a drop of water taken from an unmeasured sea."

Panthea but confirms this immensity of time in "The Vision of the Mystery." She enables him to see the sequence of life speeded up so that an age passes by in the flutter of an eyelash. He sees the changes of the earth and the changes of life corresponding. Always there is an insistence on the peculiar adaptation of each creature to its environment, on the determination of the characteristics of the organic by the conditions of the inorganic world. And always, too, "the order of life was higher, and the type of form was of a more complex character."[21] Heat turns to extreme cold. Hills wear down, continents sink beneath the sea. He sees the warfare between huge frog-like monsters, rapacious saurians and amphibia, and bird-like reptiles; "enormous animals to which the modern elephant appeared a dwarf" roam the land. "Death was the law by which the continuance of life was secured."[22]

Panthea reviews for him the succession of phenomena: the molecule of matter becomes the living cell; from the monad "races of beings, rising one above the other in the perfection of their machinery and in the order of their intelligence have arisen."[23] From these a new creation awakes—man—whose form is animal but whose mind is not from the earth but from heaven. And with the coming of man the struggle for existence ceases. "The war of races—that law of animal existence which makes the life of one dependent upon the death of another—was suspended upon the earth."[24]

Altamont's tutor confirms the teachings of Panthea, but intrudes the godly note. He too traces organic life back to the inorganic, and animal life up to man:

> . . . improving developments of form—the amorphous rock—the symmetric crystal—the organic cell—the perfect plant—the plant-like animal—the moving monad—the cyst-like creation—the invertebrate and then the vertebrate forms—the fish—the reptile—the mammal—the man—and lastly, the mind breathed into that highly organized form by the God of Creation.[25]

Like the *Vestiges* by which it must have been influenced, *Panthea* presents a highly speculative view of evolution. Here, ten years before the *Origin of Species*, is represented the law of development from the simple to the more complex, and the law of survival, the ruthless struggle for life among species. *Panthea*, too, is noteworthy as one of the earliest attempts in the novel to reconcile the orthodox view of life with the unorthodox. Thus, Panthea is made to contend that though man has evolved from the animal in form, he owes his soul or mind to God. And further, since the God who bestowed on man a soul is a benevolent god, the struggle for existence which blindly drives animals to war to the death, Panthea insists, certainly ends with the coming of man.

In the years of the nineteenth century when zoology, botany, and geology were in their infancy, the professional scientist had great use for the assistance of the intelligent amateur. The Reverend Charles Kingsley, the clergyman-novelist, was one of these. As a devotee of science in the early sense of the collecting and classifying of specimens, Kingsley not only took vast delight in such work himself, but urged it constantly upon everyone. The value that he saw in it was chiefly moral, a greater love of God through a closer knowledge of the marvels of creation.

Further, Kingsley was well acquainted with the so-called Transmutation or Development theory and with the *Vestiges of Creation* whose relation to theology he was to treat at length in his *Glaucus* (1855).[26] To this knowledge on his part we must look for the source of inspiration for the dream chapter near the end of *Alton Locke* (1850),[27] which Kingsley's friends and admirers united in condemning.[28] In that dream chapter, Alton Locke, delirious with typhus brought on by prison, heart-break, and exposure, goes through all the stages of evolution, from a madrepore at the lowest point of created life, through a crab, remora, ostrich, myledon and ape,* to primitive man.[29]

There is no explanation for this gratuitous representation of development—except that like Kingsley himself, Alton Locke had been reading many books on natural history—among them Darwin's *Voyage of the Beagle.*[30]

The Development Theory is discussed at some length in William Smith's *Thorndale* (1857),[31] a work more dissertation than novel. Like Carlyle in *Sartor Resartus,* Smith pretends to be not author but editor of the book. He tells of wandering into a villa in Naples where an

---

* In *Westward Ho!* (1885), however, Kingsley was to speak contemptuously of the "shallow and anthropologic theories" of fifty years before "that we all began as some sort of two-handed apes."

Charles Kingsley, *Westward Ho!* (Boston, n.d.), pp. 350-351.

English invalid has lately died. There he finds a manuscript containing speculations, scraps of autobiography, reports of conversations. One of these desultory dialogues turns on the animal creation and on man; whereupon the author has one character, Seckendorf by name, sound off on the various hypotheses on development. Seckendorf proceeds to discard the Lamarckian concept of the inheritance of acquired modifications arising through the influence of the environment, and favors more the suggestion broached by the *Vestiges* that development has gone on from stage to stage "by additions and varieties made in the embryo of some existing animal."

Clarence, a believer in progress, finds it difficult to account for the appearance of new creations from time to time by supposing that they have sprung full grown into being. He, too, falls back on the hypothesis of the *Vestiges* that through some environmental change which communicated itself to the embryo, some animals were pushed on "to some further stage of development."[32] But the sceptic Seckendorf hastens even further ahead of his age to philosophize over those speculations "which produce all the varied forms of organic life from some one organism —some cell, or simplest worm creeping from the hot and moist earth."[33]

In Edward Bulwer, Lord Lytton's novel *What Will He Do With It?* (1858),[34] Mr. Waite, the comedian, accompanied by his performing dog, comes upon a stationer's shop in the town of Gatesboro in which is displayed a bill advertising a "Lecture on Conchology by Professor Long." Within the shop the stationer is mechanically cutting the leaves of a two volume edition of "Researches into the Natural History of Limpets," which the mayor has just presented to the library of the Gatesboro Athenaeum. Assuming the interest of the mayor in biologic science, Mr. Waite hastens to exploit the situation to his own advantage. To the gullible mayor he offers to lay the foundation for a course of natural history "and from vertebrated mammiferes . . . gradually arrive at the nervous system of the molluscous division, and produce a sensation, by the production of a limpet."[35]

In the shop, the stationer continues to turn the pages of Professor Long's great work upon limpets, cutting those leaves which "had the volumes reached the shelves of the Library uncut, would have so remained to the crack of doom."[36]

"I do not doubt," Charles Darwin wrote in his autobiographical recollections, "that Sir E. Lytton Bulwer had me in his mind when he introduced in one of his novels a Professor Long, who had written two huge volumes on limpets."[37]

Thus, even before the *Origin of Species* (1859), the novel had manifested interest in those themes which in the middle and later Victorian age were to assume such prominence. That spleen which in time was to be visited on Darwin's works was in Disraeli's *Tancred* directed at Chambers's evolutionary hypothesis; and in Bulwer Lytton's *What Will He Do With It?* that same mirror of ridicule was held up to Darwin that soon was to serve so well for both Darwin and Huxley and the protagonists of evolution and natural selection. *Alton Locke* presents just such a rapid resumé of the course of evolution as was to feature so many Victorian compositions; and *Thorndale* adumbrates the doctrinaire novel of William H. Mallock which sidetracks the story at the expense of lengthy debate and discussion. In *Melincourt,* an acceleration of evolution brings a monkey up to man's status, and in *Nightmare Abbey* the search for a missing link is burlesqued: both themes were to recur time and again in evolutionary satire.

CHAPTER THREE

Evolution and Darwinism From 1859

Charles Darwin had waited almost twenty years to speak. When he did speak a scientific world which had scoffed at Robert Chambers's speculations and listened coldly to Herbert Spencer's *a priori* reasonings and sweeping generalizations, gave rapt attention. Evolution's day had come, and Darwin was its prophet.

With inexhaustible patience and insight Darwin brought forward in his *Origin of Species* a profound array of confirmatory facts about bees and turkey cocks, dogs and thrushes, pigeons and ants—fortifying at all points his theory of the development of species by means of variation, and the survival of the fittest through natural selection. No other theory explained the thousand puzzles which had embarrassed the old view. No other theory gripped the scientific mind with such conviction. Darwinism put life and soul into an evolutionary doctrine weak with inanition. So firm was the impress of Darwin's name and Darwin's explanation on evolution that as frequent and as common as the mistaking of Frankenstein for the Frankenstein monster is the mistaking of Darwinism for evolution.

Evolutionism is not synonymous with Darwinism. The whole immensely exceeds the part. Darwinism forms but a small chapter in the history of the far vaster and more comprehensive movement. Without that chapter, however, which explained the *how* and *why* of evolution, the scientific world had refused previously to accept the history. The eighteenth century had caught faint glimpses (as in a glass, darkly) of the descent of animals from common progenitors. With Buffon the glimpse became a distinct idea; with Erasmus Darwin the idea grew into a fully evolved and tenable hypothesis. But before the evolutionary concept in biology might commend itself generally to scientific minds a *vera causa* had to be discovered. Lamarck had an explanation to offer of the process—the suggestion that species were evolved by the transmission of acquired characteristics resulting from adaptation to environment—but he failed to convince his scientific contemporaries. Chambers had an explanation to offer—the theory suggested by Baer's discoveries in embryology that every now and then, at wide intervals of time, the foetus of some animal is advanced a step in the scale of organic perfection, and has thus become the starting point of a new and higher species—but he was condemned for treating an effect of evolution as its cause.

Darwin's contribution to the methodology of evolution was the theory of the process of natural selection.

In the public mind, however, Darwin is perhaps most commonly regarded as the discoverer and founder of the evolutionary hypothesis. It is further and more particularly believed that he was the first propounder of the theory which supposes the descent of man to be traceable from a remote and more or less monkey-like ancestor. As a matter of fact, Darwin was not the originator of either of these two cardinal theories. True, Darwin submitted a mass of evidence, carefully collected and systematically arranged, in support of the probability of the idea of evolution. But his evidence was merely the capstone of the arch built up block by block by a century of scientists. Evolutionism has been a growth of numberless minds. Laplace's name will always be held to typify the evolutionary impulse in astronomy as Lyell's in geology, Spencer's in psychology, and Darwin's in biology.

The grand idea which Darwin did really originate was not the idea of descent with modification, but the idea of natural selection—the survival of the fittest.

The principle of natural selection established by Darwin made the previously discredited notion of descent with modification immediately commend itself to the whole biological world of his time, and more particularly to the younger generation of scientists. This was Darwin's great work, his chief claim to posterity's accolade—that he made credible a theory which most people before him had thought incredible; that he discovered a tenable *modus operandi* for what before had been rather believed or surmised than definitely established. The weak point in the hypothesis of organic evolution before Darwin had been the difficulty of understanding the nature and cause of adaptation to environment. It was easy to understand by means of the clue which Darwin's discovery afforded not merely that organisms had been naturally evolved from simple primitive forms, but also the means by which they had been so evolved. Darwinism takes for granted the *fact* of evolution, and offers a tenable explanation of the *method.*

Darwin's theory of natural selection was essentially an extension to the animal and vegetable world of *laissez-faire* economics and was suggested by Thomas Malthus's theory of population. The theory rests on two main classes of facts which apply to all organized beings without exception, and which thus take rank as fundamental principles or laws. The first is the power of rapid multiplication in geometrical progression; the second, that the offspring always vary slightly from the parents, though generally very closely resembling them. From the first fact or law there follows, necessarily, a constant struggle for existence. That is to say, in every generation of every species a great many more individuals are born than can possibly survive; so that there is in consequence a perpetual battle for life going on among all plants and animals of any given generation. The majority die premature deaths. They kill each other in a thousand different ways; they starve each other by some consuming the food that others want; they are destroyed largely by the powers of nature—by cold and heat, by rain and storm, by flood and fire.

Individuals of each species differ. Some are stronger, some swifter, some more cunning. An obscure color may render concealment more easy for some; keener sight may enable others to discover prey or escape from an enemy better than their fellows. Among plants the earliest and strongest shoots may escape the slug; plants best armed with spines or hairs may escape being devoured; those whose flowers are most conspicuous may be soonest fertilized by insects. There is, therefore, both within species and as between different species, a constant competition in which the penalty of defeat is death. The weakest and least fitted to live will succumb and die, while the strongest and best fitted to live will be triumphant and survive. It is this "survival of the fittest" that Charles Darwin calls "natural selection." Nature, so to speak, selects the best individuals out of each generation to live; and not only so, but as these favored individuals transmit their favorable qualities to their offspring, according to the fixed laws of heredity, it follows that the individuals composing each successive generation are generally better suited to their surroundings than their forefathers.

Struggle and survival produce nothing new; they act only as a sieve, retaining the better organisms to perpetuate the race. Before "the fittest" can struggle and survive, it must first exist. Darwin therefore further assumes the occurrence of chance variations from the ancestral type, the causes of which, as he confessed, were unknown. When a variation occurs, enabling its possessor to survive where others die, there is a prospect of a race being formed with this peculiarity, which, slowly augmenting for thousands of generations, at last gives character to a new species. Domestic animals have been greatly changed by artificial selection; through the agency of man cows have come to yield more milk, race-horses to run faster, and sheep to yield more wool. Such facts afforded the most direct evidence available to Darwin of what selection could accomplish; and suggested to him nature's selection—natural selection improving the stock. It is true that breeders cannot turn a fish into a marsupial, or a marsupial into a monkey; but changes as great as these might be expected to occur during the countless ages established by the geologists.

Though as far back as 1838 Darwin had received the suggestion for his theory from reading Malthus's *Essay on the Principle of Population* (1798) and had sketched out the theory in some detail in 1844, so deliberate and cautious were his investigations that it was only in 1856 that he began to set in order and transcribe the evidence he had gathered together during the previous years. He was still engaged in this laborious process of literary composition in 1858 when he received a communication from A. R. Wallace, a younger scientist working in the same field, which completely upset his philosophic calm. Wallace, studying the prolific vegetation and animal life during a voyage on the River Amazon, had leaped at one bound to the same conclusion which Darwin had arrived at by years of patient research and thought—that natural selection was the process by which the evolution of species was effected. Before the Linnaean Society in 1858 Wallace's paper and a summary of Darwin's work drawn up some years before were read jointly. These innocent-looking scientific papers contained intellectual dynamite, but few realized it immediately.

The next year *The Origin of Species by Means of Natural Selection or the Preservation of Favored Races in the Struggle for Life* was published. Careless about popular esteem, Darwin remarked that he would be satisfied if the book convinced three men—Lyell, the leading geologist; Hooker, the leading botanist; and Thomas Huxley, then acquiring a reputation as a zoologist. Huxley was called upon to write the *Times* article on the *Origin.* No one then living was so well qualified by knowledge and literary ability for the task, and Huxley performed it with enthusiasm. The widespread attention commanded by the *Times* review set the stage for a controversy on evolution which raged furiously for about twenty years, conducted vigorously — and sometimes bitterly — with tongue and pen. In the drawing room the *Origin of Species* competed with the last new novel, and in the study it was troubling alike the man of science, the moralist, and the theologian. On every side it raised a storm of mingled wrath, wonder, and admiration.

The scientific camp was divided immediately, the line of cleavage following closely an age-distinction. The older men, who had lost the elasticity of judgment so necessary to the reception of new concepts, refused to shake their views of the eternal immutability of species:[1] the veteran zoologist Owen, resolutely hostile, the geologist Lyell, not wholly convinced. The younger spirits, the men of the future, headed by Huxley, were full of enthusiastic faith. By these last Darwin was hailed as a second Newton; his *Origin,* as a second *Principia.*

In the *Origin of Species* Darwin did not attempt to apply the theory of natural selection to man, although it was evident that any law which affected the whole organic world must necessarily include the highest member of it. He chose rather to avoid the issue, while realizing that by his work "light would be thrown on the origin of man and his history."[2] But Darwin's cautiousness in this regard was not shared by others. Almost from the publication of the *Origin of Species,* the ground was shifted to the question of man's relation to the other animals.

Two years before the *Origin of Species* appeared, Huxley had begun investigating the structural relations of apes and men. He was inspired to this task by the conviction that the pronunciamentos on this subject of Sir Richard Owen, the greatest living comparative anatomist,* were in flat contradiction with the facts. Fortified by his own researches, Huxley was, however, not altogether ready when in June 1860 at a meeting of the British Association for the Advancement of Science, Owen decided to parade his views. With the deliberate intention of employing all his great prestige as the "British Cuvier" to discredit Darwin, Owen prepared a paper on the comparative anatomy of the brain with especial reference to the differences between the human and anthropoid brains.

The esteemed biologist presented his facts. They allowed of no other interpretation, he averred, than that the anthropoid brain is radically different from the human brain. Specifically, it contains no *hippocampus minor.* While a certain vague and general resemblance might perhaps be observed between the human and other brains, the gap between the human brain and that of the highest anthropoid is greater than the difference between the brain of the highest ape and that of the lowest monkey.

Now, this is not true; and if there was anyone who knew it was not true, it was Huxley. He had already plowed through this field, thoroughly. The public effect of such statements uttered with absolute finality might be most unfortunate, Huxley felt. He had, however, prepared no

* William M. Thackeray pays due tribute to Professor Owen's skill—and his own—in a passage in *The Newcomes* (1853-55) identifying the novelist's task with that of the comparative anatomist's:

How can I tell the feelings in a young lady's mind; the thoughts in a young gentleman's bosom?—As Professor Owen or Professor Agassiz takes a fragment of a bone, and builds an enormous forgotten monster out of it, wallowing in primaeval quagmires tearing down leaves and branches of plants that flourished thousands of years ago, and perhaps may be coal by this time—so the novelist puts this and that together . . .
William M. Thackeray, *The Newcomes* (Boston, 1891), II, 340.

paper in reply and had no materials for demonstration. Therefore he did a most unusual thing: he gave Professor Owen the "lie direct." On public record Huxley placed a deliberate and unqualified contradiction of Owen's "facts," at the same time pledging himself to justify that procedure elsewhere.

In January, 1861 Huxley redeemed his promise to Owen with a crushingly complete memoir in the *Natural History Review;* and in 1863 in his first book, *Evidence as to Man's Place in Nature,* Huxley boldly stepped to the front, publicly applying the Darwinian theory of natural selection to the human race, and proving beyond contention that man could not by virtue of any exceptional physiological peculiarities claim a special position outside of or above all other organisms.

Man's Place in Nature was a deliberate flouting of popular superstition. Of all the burning questions connected with the origin of species, this was the most heated—the most surrounded by prejudice and passion. To touch it was to court attack, to be exposed to endless scorn, ridicule, misrepresentation, abuse—almost to social ostracism. But the facts were there, and Huxley was prepared to go through fire and brimstone for those facts.

There is no doubting the relish with which Huxley detailed the characteristics of the anthropoid ape to a mid-Victorian public which had never seen the like; for when Huxley wrote there was no such thing as a live ape in a London "zoo." He described the great ape's power and ferocity, then went on to present the structural evidence of man's relation to the anthropoids, beginning with the embryo and concluding with the brain and a note on the famous *hippocampus* controversy with Owen. The last section treated of fossil man, exhibiting the links such as Neanderthal man spanning the short gap between ape and man.

It was this last section treating of the structural interval which exists between man and the man-like apes which started the now-famed career of the "missing link." Huxley had brought forward all the available evidence, but he had to admit the gaps in the fossil record. If one group of animals has been derived by descent from another, it is obvious that there should be some species more or less intermediate between the two groups—species exhibiting the characteristic features of both. Thus, among vertebrates, examples of such a connecting link were available in those extraordinary fishes known as lung-fishes that have both the gills of fishes and the lungs of land vertebrates. But no such fossils, living or dead, were on hand to fix beyond doubt man's place in nature. Owing to the vastness of geologic time, to the age-long destruction of fossil evidence by erosion, and to the limited number of those who collected fossils, evolutionists of Huxley's day had to lean heavily on the hypothetical connecting links based on studies in comparative anatomy and embryology; they had few concrete exhibits. Huxley, however, was optimistic. Who knows but that in some geological stratum, he queried, "the fossilized bones of an ape more anthropoid, or a Man more pithecoid, than any yet known await the researches of some unborn paleontologist?"[3]

Opponents of his theory insisted that he produce such a link, this so-called original ancestor of man and the anthropoids from which the species branched off, this creature supposedly combining simian and human qualities. Further, they would remain incredulous until such time as he was prepared to present as evidence fossil remains leading gradually down from man or up from the ape.

It was this hypothetical pithecoid progenitor of the race of man and of the great apes that soon came to be distinctively styled "The Missing Link"; and the equally hypothetical stages assumed to intervene in evolution between the ape and man were labeled generically "connecting links" or "missing links."

Simultaneously with Huxley's *Man's Place in Nature* appeared Sir Charles Lyell's *Antiquity of Man.* Sir Charles, the most eminent of living geologists, a man of deeply Christian feeling and of exceedingly cautious temper, had opposed the evolution theory and declared his adherence to the idea of successive creations. But in the *Antiquity of Man* he showed himself a complete though unwilling convert to the fundamental ideas of Darwin. The support of Lyell was not unexpected. In various review articles against the Darwinian theory there had been appeals to Lyell, at times almost piteous, not to flinch from the truths he had formerly proclaimed. On the other side, Darwin and his scientific cohorts had been bombarding Lyell with every piece of new evidence that had been adduced in support of the theory. The aging geologist yielded unreservedly to the mass of new proofs arranged on the side of evolution against that of creation. His decision was tremendously important to the social success of evolution.

Lyell's work was at once a consummation and a beginning. The palaeontological work of Agassiz and Owen, the accumulations of evidence from drift and bone cave, from barrow and picture writing, had been forcing upon the minds of both antiquaries and scientists, already convinced of the antiquity of the world, the belief that man, instead of dating back a mere trifle of six thousand years or so, was really contemporary with the mammoth, the cave bear, and other extinct quaternary animals. To such research as Huxley was carrying on in applying the theory of natural selection to man, Lyell's epochal treatise was but an introduction. So long as it was believed that man had been on earth only six thousand years it was impossible to apply to him a theory involving variations in species whose development would require aeons of time. To provide for the gradual evolution of species by means of natural selection it had been necessary first to demonstrate the hoary antiquity of the earth: geologists and palaeontologists did this. To provide for the emergence of man from lower forms under the same laws, required extension backwards of man's origin to times infinitely remote: Lyell's *Antiquity of Man* established this.

The success of the *Antiquity* and *Man's Place* was immediate, despite such criticisms as that of the *Athenaeum* that "Lyell's object is to make man old, Huxley's to degrade him . . . man probably lived a hundred thousand years ago, according to Lyell; man probably had a hundred thousand apes for his ancestors, according to Huxley." These works and that of Darwin were torn from the hands of Mudie's shopmen as if they were novels.[4] Resultantly, during the ensuing decade, hard upon the heels of the volumes of Huxley and Lyell, followed quite a library of books concerning the genealogy of man, culminating in 1871 in Darwin's *Descent of Man.* Sir John Lubbock's *Prehistoric Times* appeared in 1864; his *Origin of Civilization* in 1870. And in 1868 Ernst Haeckel, who in continental Europe was doing for Darwinian evolution what Thomas Huxley was doing in the English speaking world, published his *History of Creation.*

The publication of Huxley's *Man's Place in Nature* and Haeckel's *Creation* showing the inevitable extension of the Darwinian theory to the human species, forced Darwin's hand. By his *Descent of Man* the author of the *Origin of Species* accepted the responsibility for this extension which he had recognized from the first. The *Descent of Man* was a judicial summing up of the entire question of man's origin in which Darwin dealt also with the process of selection in relation to sex.

The evidences "that man is descended from a hairy, tailed quadruped, probably arboreal in its habits" collected and marshaled by Darwin, consist of minute inferential proofs of similarity of structure, at certain stages of development, between man and the lower animals. This similarity is especially marked in the embryonic stages; and taken with the existence in man of various rudimentary organs like the animal ear-point, the vermiform appendix, the general hairiness of the human body, and the hidden little bony tail of man—parts left behind as the result of change of life-habit or environment—seems to imply that man and the lower animals come from a common ancestor. "The Quadrumana and all the higher mammals are probably derived from an ancient marsupial animal and this through a long line of diversified forms, from some amphibian-like creature, and this again from some fish-like animal."[5]

Darwin taught that the mind of man in its lowest stages was essentially an animal mind. The result of his investigations was to show that, great as the difference in mental powers between man and higher animals may be, it is undoubtedly only a difference "of degree and not of kind." The upward progress of mind is viewed as effected by natural causes. Just as the bodily organism is capable of varying in an indefinite number of ways, so may the mental faculties vary indefinitely with certain physical changes. In this way Darwin sought to account for all the higher mental powers usually regarded as absolutely peculiar to men, like the use of language and reason, the sense of beauty, and conscience.

Darwin reasons that the early ancestors of man must have been more or less monkey-like animals of the great anthropoid group, and related to the progenitors of the orang-outang, the chimpanzee, and the gorilla. Opponents of evolution often refer to this derisively as "the monkey theory" of man's descent. Darwin, however, never maintained that man is a lineal descendant from the ape or monkey, but rather that men and apes long ages ago branched off from a common ancestor of anthropoid family.

Precisely the same causes are suggested for the origin of man as for the origin of species in general. Variability occurs in man to a high degree; and those individuals who, by chance variation, depart profitably in one way or another from their parents are best fitted to succeed. Moreover, the rapid multiplication of the human race creates conditions which necessitate an energetic struggle for existence in which those that are least capable of adjusting themselves to the conditions of existence are crowded out and perish, leaving the field to the more capable, the fittest to survive. The fittest, by surviving, raise the general level of the race. The weak die early or bear no children, and the children of the strong inherit the superior qualities of their parents.

Associated with this theory of natural selection Darwin also advanced a theory to account for those often marked differences between the two sexes of a species (other than those of the reproductive organs themselves), which are known as secondary sexual characters. These include special organs of attack and defense used by rival males in fighting for the possession of females, and the wonderful developments in plumage, conspicuous colors and pattern, the latter of which would seem to be elements of disadvantage rather than of advantage in the struggle for life. Darwin's explanation of this, called the theory of sexual selection, was that these characters are of advantage in the rivalry of mating, many of them being apparently of a kind to attract and stimulate individuals of the opposite sex. Hence they might help the possessors win in the struggle to find mates and consequently to leave progeny. Darwin is inclined to attribute the nakedness of man, then, not to the action of natural selection on ancestors who originally inhabited a tropical land, but to sexual selection, which, for aesthetic reasons, brought about the loss of hairy covering in man, or primarily in woman. Sexual selection may be considered as a subordinate and special kind of natural selection, not involving a determination between life and death, but one between going childless and leaving posterity—which is, after all, the essential determination in natural selection.

To the demonstration and elaboration of his theory, Darwin's later years were devoted. He pursued a series of studies, the results of which appeared chiefly in the following books: *The Fertilisation of Orchids* (1862), supplemented by *Cross and Self-Fertilisation* (1876). *The Movements and Habits of Climbing Plants* (1864); *The Expression of the Emotions in Man and Animals* (1872); *Insectivorous Plants* (1875); *The Power of Movement in Plants* (1880) and *Formation of Vegetable Mould through the Action of Worms* (1881). He died in 1882.

The establishment of the theory of evolution was the great intellectual event of the century; indeed, one must go back to the substitution of the Copernican for the Ptolemaic system to find a parallel. Within twenty years even the divines had become reconciled to the teaching of evolution given in 1859. But within twenty years the biologists had begun to evince, ironically enough, a growing scepticism about some of the Darwinian conclusions which had established "descent with modification" universally as the order of nature in the organic world. In particular the principle of natural selection—Darwin's mechanism for the process of evolution—was subjected to critical scrutiny.

Darwin himself was under no illusion as to the omnipotence of Selection. At the close of the Introduction to his *Origin of Species* he qualified his conclusions by remarking, "I am convinced that natural selection has been the main but not the exclusive means of modification."[6] The whole body of scientific opinion followed Darwin in this, holding that while natural selection was the "main factor" of evolution, nevertheless it was largely supplemented in its work by certain other subordinate factors. Of these the most important were taken to be the inherited effects of use and disuse, together with the influence of the environment in directly producing alterations both of structure and of instinct. This was Lamarckism come to life once more, a theory of evolution which Darwin, though disparaging, had accepted in some measure. Lamarck, it will be remembered, had held that the giraffe has got a very long straight neck because of the cumulative result of generation after generation of stretching up to the branches of the acacia-tree; a cave animal is blind as the result of ages of living in darkness, during which the eyes have suffered from disuse. With certain provisos Darwin inclined to accept this view as supplementary to his own. The neo-Lamarckians following him were about ready, on the other hand, to scrap the natural selection hypothesis of Darwin for Lamarckism *redivivus*.

Shortly after Darwin's death, however, this state of matters underwent a very serious change. For it was just after 1882 that Professor August Weissman began to publish a remarkable series of papers whose primary object was to establish a new theory of heredity which had for one of its consequences a denial of the inherited effects of use and disuse or indeed of any characters which are acquired during the life time of individuals. According to this theory which out-darwined Darwin, natural selection is constituted the sole and only cause of organic evolution; the only kind of variations that can be transmitted to progeny are those which are called congenital. Weissmann changed the fashion of science and directed its attention to the chemico-physical processes of life by expounding the theory of the continuity of the germ-plasm. By this theory a distinction is made between body plasm and reproductive plasm. While most of the material of the fertilized ovum is used to build up the body of the offspring, its nerve and muscle, blood and bone, a residue is kept intact and unspecialized to form the beginning of the reproductive organs of the offspring, whence will be launched in due course another organism on a similar voyage of life. Thus the parent is rather the trustee of the germ-plasm than the producer of the child; the age-old question whether the hen made the egg or the egg made the hen, was answered: the fertilized egg makes the hen and the eggs thereof. As Professor Bergson puts it, "Life is like a current passing from germ to germ through the medium of a developed organism."

The exclusion of Lamarckian factors from evolution was distasteful to many competent naturalists who continued to insist that natural selection was insufficient to explain the origin of species. Its exclusion likewise found no favor with many writers, whose chief interest in biology was its application to man, and to whom it seemed incredible that the influences of education and civilization affected only the generation which experienced them. But in time this objection was removed when it was realized that evolutionary development has, in the human species, been transferred from organic elements physiologically inherited to social tradition: The difference between savagedom and civilization is not organic, but cultural.[7]

Another great change that has come about since Darwin's day is a recognition of the frequency of discontinuous variations, by which is meant sudden novelties which are not connected with the type of the species by intermediate gradations. Especially through the investigations of Professor Hugo de Vries it has become plain that changes of considerable magnitude may occur at a bound. When the new character that suddenly appears, such as a short-legged Ancon sheep, has a considerable degree of perfection from its first appearance, and reappears persistently and intact in a certain proportion of the descendants, it is called a *mutation*. Darwin assumed the occurrence of fluctuating variations which are marshaled before the tribunal of natural selection which passes upon them the sentence of death or survival; i.e., species by gradual modification are slowly changed into new types. The mutation theory of De Vries, however, explains the origin of species by sudden and saltatory leaps rather than by gradual modification. *Natura facit per saltum.* Post-Darwinian research by experimental evolutionists disclosed that variations are not infinitesimal, but are in general rather of the nature of mutations. Resultantly, the vast periods which were formerly believed to be essential for the differentiation of a species were shown to be far too large.

The work of Weissmann and De Vries, of Pavlov, Galton, and Mendel since Darwin's time has shown that we inherit from our parents not noses and eyes as we inherit real estate or money—the belief in Darwin's time—but genes, units of life, in the cell; and in this direction subsequent research has tacked. No longer by observation and inference alone, but by experiment with these factors, evolutionary scientists seek to explain how evolution has come about. Through the research of experimental evolutionists significant changes have taken place in the concept of these *factors* at work in evolution; in the *fact* of evolution, none. While the majority of scientists today no longer admit the sufficiency of natural selection alone to explain evolution, and may be said to have discarded Darwinism, there is probably no scientist of repute in the world today who does not admit the fact of evolution in the sense of a progressive development to higher and more complex forms of life.

CHAPTER FOUR

THEOLOGY VERSUS EVOLUTION AND DARWINISM

It is difficult for us today, almost four generations removed from the age of Darwin, to conceive of the stormy controversy that raged then over his cataclysmic discoveries. The growth of the scientific temper within recent years has secured a fair reception for any new doctrines, however unpalatable at first sight. But in the 'sixties, 'seventies, and 'eighties such were the fears of faith in the presence of a godless science that the mere threat of a new scientific idea was enough to flutter the clerical dovecotes.

Twice before the Church had faced Science in mortal combat for the possession of the world, and twice the Church had lost the world. The first great blow which Science inflicted was that which destroyed the old geocentric astronomy, and showed that this world is "but a sand-grain in the continent of Being." The next great blow came from modern geology which was felt at once as a menace to the doctrine of direct creation of the world. Bitterly the Church fought each losing battle, smiting the infidel with every weapon in its armory. The Church learned nothing from its two defeats. Unlike science which sets up hypotheses and then begins at once trying to disprove or modify them, religion, its hypotheses set, defends them to the last syllable whatever the cost.

The *Origin of Species* came into the theological world like a plough into an ant-hill. Everywhere those rudely awakened from their old comfort and repose swarmed forth angry and confused. The Church factions—all of whom believed implicitly at that time in the literal truth and inspiration of the Biblical story of the Creation and Fall of Man—set up a hue and cry: "Darwinism seeks to dethrone God!" Reviews, sermons, books light and heavy came flying at the new thinker from all sides.

Victorian decency had been tactfully closing its eyes to the havoc that German scholarship had already made with the literal interpretation not only of the Old, but also of the New Testament. Theologians of every sect and creed chose persistently to blink the facts of geological science and teach that only some six thousand years had elapsed since the earth sprang into being. Everywhere it continued to be accepted that the innumerable species of animal and vegetable as geology reveals or nature displays them, were separately and divinely created. Secure in its complacency, even at the height of the Victorian noontide, the age persisted in ignoring the ominous fissures and subsidences which had begun to make themselves apparent and, as the century turned the half mark, became wider and deeper.

Here and there in this period of intellectual ferment were individuals who with ear to the ground could hear the rumblings of the temblor that was about to shake the theological and scientific world to its foundations. One of these was the father of Edmund Gosse, Philip Henry Gosse the naturalist, who with pathetic devoutness rushed to save orthodoxy by filling up the widening crevices with a theory justifying geology to godly readers of Genesis. Gosse argued that there had been no gradual modification of the surface of the earth or slow development of organic forms, but that when the sudden act of creation took place, the world presented instantly the structural appearance of a planet on which life had long existed. Everything was constructed *as if* it had a past history; just as Adam and Eve (so theologians had decided) were created with navels as if they had been born in the ordinary way. Significantly, Gosse called his book *Omphalos* (navel, Greek). The rocks, Gosse held, had been filled with fossils and had been made just such as they would have become if they had been due to volcanic action or to sedimentary deposits. To his bitter disappointment nobody could believe his logically admirable reconciliation of theology with the data of science.

Gosse's attempt to fling geology into the arms of Scripture was born of a pious belief, which he shared with the majority of his contemporaries, that the Bible as Revelation was infallible, as infallible in its science as in its ethics. To such an outlook, the discovery of the process of organic evolution delivered a staggering blow, as fatal to the doctrine of the creation of man as geology was to the doctrine of the direct creation of worlds. Darwin's theory gave the "final push to the already shaky throne which the worshippers of the Lord had chosen to erect for him on a pile of Family Bibles."[1] It became obvious to anybody who chose to think that the seven day's creation and the Adam and Eve story were on a par of literal veracity with Jack the Giant Killer.

The implications of a purely naturalistic interpretation of life were bound to be felt by many people as a shock to their faith. Theologians were quick to point out from the very publication of the *Origin of Species* the challenge to orthodoxy implicit in the Darwinian hypothesis. They went even further than Darwin was prepared to go just then in suggesting that the theory of the mutability of species could be easily extended to support the lineal descent of the human race from the lower animals. The conception of a vast system of physical causation, embracing man and nature in one iron whole, seemed to leave no room for religion.

Darwinism was charged not only with the negation of religion, but also with hunting God out of the world. The Darwinian principle of natural selection was considered by many as having dealt a fatal blow to the validity of the teleological or design argument for the existence of a Supreme Intelligence in the universe. Coming at a time when Paley's classic statement of the argument from design in nature enjoyed well nigh universal acceptance, the Darwinian explanation seemed to take the foundation completely from under it. Thus, Paley has reasoned that if a passer-by found a watch lying in the sand, he would conclude from the ingenious arrangement of its parts, designed to measure the passage of time as checked off by the stars, that the watch did not just happen that way but had been made by an intelligent watchmaker. So likewise the intricate adjustments and delicate coordinations that are apparent in nature bear irrefutable evidence of the work of a Supreme Architect or Designer.

But Paley and his age, still in the afterglow of Newtonian physics, conceived of the world as a cosmic machine capable of running itself except for a little jog or poke here and there in the shape of a special Providence: in brief, a world made. But for this conception of a "world made," evolution substituted that of a "world making"; the watch is still being adjusted. But far from following any design or pattern, the adjustments or adaptations according to the Darwinian theory of evolution are spontaneous and fortuitous, accidental variations occurring in all directions and in all degrees, from which the rough ordeal of the struggle for existence picks out and perpetuates those best adapted to the environment and best fitted to thrive in it. They simply happen, grow, and are preserved. In consequence of this evolutionary view, Paley and his watch were consigned to the scrapheap; and the idea of design or purpose in the organization of the universe and in the operations of nature for a time had to be surrendered. Nor could the host of arguments for the beneficence of Providence, derived from the exquisite adaptation of animals to the environment, hold water, for incompatible with divine beneficence was the wastefulness and cruelty in nature suggested by Darwin's emphasis on the struggle for survival.

But worse than these, it was held, was the affirmation that man was descended from the lower animals. Theologians and uneducated people, indeed, fastened upon this one aspect of the theory and upon it burst the full force of the anathematizing *odium theologicum.* "Darwin says that men are descended from monkeys!" the world exclaimed in horror. The principal obstacle which the doctrine of evolution encountered in the popular mind was that the conception of man being the lineal descendant of "ancestral apes or patriarchal pollywogs" is degrading to man's dignity. Man had been regarded as the aristocrat of the universe, as the crowning work of the Creator. Here it was shown that he was only an animal among others; democratizing science now taught him not only his equality but also his fraternity with the ants and apes. One did not have to be an orthodox Christian to object to what seemed an assault on human uniqueness and superiority. Consequently, there was a very general snarling and arching of backs at this theory of man's essential bestiality.

Not for the first time was the world ready to scrap a significant discovery deemed lacking in decorum. When sunspots were first observed they were indignantly denied by the Aristotelians, on the ground of its being impossible that the eye of the universe could suffer from ophthalmia; and when Kepler made his great discovery of the accelerated and retarded motion of the planets in different parts of their orbits, many persons refused to entertain the conception, on the ground that it was "undignified" for heavenly bodies to hurry and slacken their pace in accordance with Kepler's Law. The progress of science seems destined rudely to dispel any ideas of dignity whether of human bodies or of heavenly bodies whenever these ideas have been applied to her processes.

The theologians saw what was involved more clearly than did the general public. If man is to be placed wholly within the organic realm, doubts as to the unique spiritual nature of man immediately arise; if man developed by imperceptible steps from animal ancestry at what moment did he acquire his immortal soul? Sin or carnal tendencies, in the evolutionary view, derive not from the Fall, but are wholly to be explained by the natural survival in man of animal propensities, inherited from brute ancestors. And if the idea of man as the accidental product of blind, physical processes was to be believed, there was no gainsaying the harm that would be done to religion by the implied encouragement to fatalistic and hedonistic philosophies.

Religion girded for war against biology for the soul of man as it had before against astronomy and geology for the kingdom of man.

The lead was at once taken by the then Bishop of Oxford, Samuel Wilberforce, a divine with some pretensions to scientific knowledge. He was in his day the foremost member of his own profession, by general consent the most effective preacher in the Church of England. "Soapy Sam's" (Wilberforce) own knowledge was inadequate; but after thorough coaching by Sir Richard Owen, *doyen* of British biologists with whom Huxley had crossed swords on the *hippocampus minor* question, he set to work on two fronts to smash Darwin.

The first attack was in the form of an article in the *Quaterly Review* of July, 1860 denouncing the principle of natural selection as incompatible with the word of God. But even before this was off the press, the Bishop made ready for a second sortie—this time an address at the annual meeting of the British Association for the Advancement of Science, held in Oxford. A tacit understanding had been reached between the evolutionists and the anti-evolutionists for a pitched battle; Dr. Wilberforce, again primed by Professor Owen, was to act as mouthpiece to the great palaeontologist, and open the debate on behalf of the Church. To demolish Darwin the prelate selected the theory of the descent of man and the apes from a common ancestor as his main ground of attack.

Two days before, in "Section D," the zoological division of the association, Thomas Henry Huxley had openly contradicted Sir Richard Owen's pronouncement on the structural differences of man and ape. Now at the earnest solicitation of his friend Robert Chambers, the reputed author of the *Vestiges of Creation,* Huxley agreed to appear in this climactic battle which was to loom large in the public eye because it was not merely the confounding of one anatomist by another, but the open clash between Science and the Church.[2]

Darwin had no taste or talents for controversy, and little concern for the rapid diffusion of his doctrines; he cared only to pursue his researches recognized by the expert few. Huxley, on the other hand, a man whose controversial complexes suffered no repression, was only too ready to constitute himself "something between maid-of-all-work and gladiator-general for science." Armed cap-a-pie he jumped into the evolution controversy.

He did not enter the lists altogether unknown like the Disinherited Knight of Scott's immortal romance. At twenty-five he had gained a fellowship of the Royal Society after several years voyaging, like Darwin, to far places. Soon after, he was being treated as an equal by men of established reputation. An evolutionist before the *Origin of Species,* on its publication Huxley assured Darwin that he was ready, with claws and beak, to sustain the attack that was sure to come.

With florid and fluent rhetoric the prelate assured his audience that there was nothing in the idea of evolution. "What have they to bring forward?" he exclaimed. "Some rumoured statement about a long-legged sheep." Then turning to Huxley with a smiling insolence, he begged to know, was it through his grandfather or his grandmother that he claimed his descent from a monkey?[3]

This was a fatal mistake. Huxley instantly grasped the tactical advantage which this *ad hominem* argument of the Bishop's gave him. He turned to Sir Benjamin Brodie who was sitting beside him on the platform, and confided, "The Lord hath delivered him into my hands." The bishop finished his speech, and the audience called for Huxley. What Darwin's "bulldog" then proceeded to say no two witnesses exactly agree, but on this they are all one, that then and there Huxley won his spurs as a dangerous adversary in debate and a personal force in the world of science which could not be neglected. After setting the prelate right in his biology by explaining that the suggestion was of descent through thousands of generations from a common ancestor, Huxley went on to make his famous peroration:

> I asserted—and I repeat (John Richard Green quoted him as replying to the hapless bishop) that a man has no reason to be ashamed of having an ape for his grandfather. If there were an ancestor whom I should feel shame in recalling it would rather be a man . . . who . . . plunges into scientific questions with which he has no real acquaintance, only to obscure them by aimless rhetoric . . . and skilled appeals to religious prejudice.[4]

With incomparable éclat Huxley had spitted the bishop. Poor Wilberforce's rhetoric was burst like a paper bag. But the real drama behind this exchange of personalities was entirely lost on the audience. Huxley had done much more than make a monkey of a bishop. He had saved a great cause from being stifled under misrepresentation and ridicule. Instead of being crushed under sneers, the new theories secured a hearing, all the wider indeed for the dramatic nature of their defence.

But the Church was far from through. It had only just begun to fight. Catholics and Protestants alike surged forward to do battle against the infidel, a troop in chain armor and with cross-bows on a nineteenth century battlefield. The Church sought to undermine the evolutionary argument by founding sacro-scientific organizations. To combat "science falsely so-called," the Victoria Institute was established by the Protestants; and the Academia, planned by Cardinal Wiseman, was designed by the Roman Church. In a circular letter, the cardinal, usually so moderate, pontificated, "Now it is for the Church, which alone possesses divine certainty and divine discernment, to place itself at once in the front of a movement which threatens even the fragmentary remains of Christian belief in England." And as a sort of antiphonal response Cardinal Manning, leader of the English Catholics, declared his abhorrence of the new view of nature, and described it as "a brutal philosophy—to wit, there is no God, and the ape is our Adam."[5]

These attacks from such eminent sources set the clerical fashion for several years. As from a thousand laboratories facts poured out substantiating the evolutionary concept, from a thousand pulpits anathema was hurled on the ranks of its protagonists. Evolution was one thing, Darwinism still another, and naturalistic philosophy a third entity; but outraged churchmen lumped together, confused, and confounded all three, damning them as scientific heresies to be condemned rather than seriously discussed. Science, they thundered, robbed the world of its poetry and spirituality. It interpreted existence in terms of mechanism, consigning the whole Christian cosmogony to the rubbish heap. Especially were the vials of contempt poured over Huxley, Tyndall, Spencer, and Darwin, the four Evangelists of the Gospel of Unbelief, as the *Quarterly Review* called them. Here were men who would "peep and botanize" on a mother's grave.

Especially were they exercised over the attack on Revelation. The Holy Writ had been under attack for some time from the critical side, and Bishop Colenso's *Examination of the Pentateuch* (1862-1879), proving these books post-exile forgeries, was even then riddling the authenticity and literal truth of the Mosaic cosmogony. This was bad enough, but so long as Biblical criticism was confined to the genuineness of certain books, the interpretation of prophecies, and the determination of dates, the great body of the laity were only feebly interested in the struggle, and the churchmen only slightly alarmed. It required a vast deal of scholarship which few but the controversialists possessed, to understand the points of controversy and their precise bearings. But the doctrine of evolution was not a point of scholarship but a law of nature, and every newspaper and magazine had made of its every element grist for the journalistic mill. Consequently, not merely trained scholars who had spent a lifetime upon the niceties of verbal criticism, but the great mass of the more intelligent laity took part in the controversy. Frightened, the Church drew its lines closer around Scriptures.

But in 1863 came an event which brought serious confusion to the theological camp—the publication of the *Antiquity of Man.* Sir Charles Lyell had gone over completely to the enemy. The veteran geologist, for long a pillar of the anti-transmutationists, had hesitated so long in espousing the Darwinian hypothesis that the theological contingent had been inspired to hope they had in him an ally. Nine editions of the *Principles of Geology* had appeared, and in every one of them Lyell had maintained the doctrine of Special Creation. But he who had really paved the way for the evolutionary doctrine by his own proof of geological uniformity finally pushed sentiment aside and declared himself a Darwinian. The blow to theology was serious in many ways, and especially so in two—first, as withdrawing all foundation in fact from the scriptural chronology, and secondly, as discrediting the Special Creation theory.

At the same time came Huxley's *Man's Place in Nature* nailing the articles of evolution to the doors of orthodoxy, shortly to be followed by Ernest Haeckel's *Schöpfungsgeschichte.* The interpretation of evolution and its popularization for the masses by Huxley in England and by the Jena professor in Germany only served to intensify the fears of the religious minded. Running through the Grand Elucidator's interpretation there was a generous vein of agnosticism, while in that of Haeckel there was a strident and blatant materialism which was even more offensive to religious ears. Under the circumstances orthodox folk came to look upon Darwin as a near-relative of anti-Christ, and well-meaning men who knew nothing of evolution except that it was a theory of natural history unsanctioned by Scripture, shuddered at those who were endeavoring to advance science as "atheists" and "unbelievers." Thomas Carlyle, who preserved the intolerance of the orthodox without their creed, was one of these. Soured and embittered, his once supple mind now long closed to new ideas, the sage of Cheyne Row did not forgive the publication of *Man's Place in Nature.* Years after, near the end of his life, Huxley saw him walking slowly and alone down the opposite side of the street, and touched by his solitary appearance, crossed over and spoke to him. The old man looked at him, and merely remarking, "You're Huxley, aren't you? the man that says we are all descended from monkeys," went on his way.[6]

In spite, therefore, of many successes on scientific fronts Darwin's new evolutionary creed, faced by obscurantist tactics and appeals to religious prejudice, had still to struggle for existence.

Bishop Wilberforce, using just such a claptrap platform appeal had to retire in disorder before the big guns of Huxley at the Oxford meeting of the British Association in 1860. But in the autumn of 1864 at Oxford again, the Bishop struck back, in behalf of the Church, this time through an ally. On the direct invitation of Wilberforce, Benjamin Disraeli—then leader of the Opposition in the House of Commons—was asked to make a "great speech" at a meeting held in the Sheldonian theater at Oxford.

As a Conservative Disraeli execrated the critical principle which inspired latitudinarians and evolutionists alike to call in question Semitic Scriptures. Both of these infringed on the monopoly of Moses. Here was a great opportunity to set forth his own ideas and hold up the newfangled to scorn and derision. With that superb nonchalance in which he was unrivalled among the orators of the day, the Opposition whip proceeded first to pulverize the Broad Church party of the day, then turning against the evolutionists, concentrated in an unforgettable phrase his most vital convictions: "What," he asked, "is the question now placed before society with glib assurance the most astounding? The question is this—is man an ape or an angel? My Lord, I am on the side of the angels."[7]

The victory was a hollow one, as sophistical as Johnson's triumph over Berkeley and Idealism with the toe of his boot. Darwinism in 1864 was far beyond the point where it could be checked by an epigram. *Punch* represented "Dizzy" dressed for an Oxford *bal masque,* equipped with enormous wings, in a beatific pose before a mirror.

Huxley came up against orthodoxy again with his Edinburgh lecture in 1869, *On the Physical Basis of Life.* In this he contended that the living body of plants and animals is made up of similar material, and that vital action and even thought are ultimately based upon molecular changes in this life-stuff. This "protoplasm," Huxley declared, some thought he had invented for the special purpose of plaguing the orthodox. But five years later his view was taken up and reinforced by Professor W. K. Clifford who interpreted from it the possible existence of consciousness apart from the nervous system. By that time John Tyndall, too, had delivered at Belfast the presidential address to the British Association. In this he justified the derivation of life from spontaneous combination of its inorganic elements, and ruled out the necessity of imagining any other cause. "We claim," he said, "and we shall wrest from theology the entire domain of cosmological theory."

In 1871 was published Darwin's *Descent of Man.* Its doctrine had been anticipated by critics of his previous books, but it made none the less a great stir; again the opposing army trooped forth as in crusade, but evidently with much less heart than before. Yet again Darwin was accused of being resolved to hunt God out of the world. In an address at Liverpool, Mr. Gladstone remarked: "Upon the grounds of what is termed evolution God is relieved of the labour of creation; in the name of unchangeable laws he is discharged from governing the world." Herbert Spencer pointedly called his attention to the fact that Newton's law of gravitation and the science of physical astronomy are open to the same charge.[9] Carlyle spoke petulantly of Darwin's doctrine of descent from ape-like ancestors as "rather a humiliating discovery, and the less said about it the better."[10]

But it was noted that this second series of attacks, on the *Descent of Man,* differed in one remarkable respect —so far as England was concerned—from those which had been made over ten years before on the *Origin of Species.* While everything was done to discredit Darwin, to pour contempt upon him and even to make him "a persecutor of Christianity," while his followers were represented as charlatans or dupes, there began to be in the most influential quarters careful avoidance of the old argument that evolution—even by natural selection—contradicts Scripture. It began to be felt that this was dangerous ground. The defection of Lyell had, perhaps more than anything else, started the question among theologians who had preserved some equanimity, "What if, after all, the Darwinian theory should prove to be true?" Recollections of the position in which the Roman Church found itself after the establishment of the doctrines of Copernicus and Galileo naturally came into the minds of the more thoughtful. The war of Genesis versus Evolution passed into a new phase.

CHAPTER FIVE

SATIRES ON EVOLUTION AND THE 'MONKEY THEORY'

The battle of the century between Natural Selection and Religion was by no means a private fight, closed to the public. At first, it was true, Darwinism had to run the gantlet unaided between the armed files of orthodox churchmen. But the contest ceased to be an exclusive one between Science and Church. It became in fact a battle royal as soon as the issues of the struggle, spread out on the pages of every journal of the day—in the sedate columns of the *Times*, on the frolicking sheets of *Punch*—began to vie for the attention of *mobile vulgus* with the latest Reform Bill, the Civil War in America, the Home Rule movement in Ireland. The Victorian novel, ever alive to the electric impulses generated by the age, prepared to transform this new controversy into the matter of fiction. This was inevitable. Here was a subject of universal import that struck to the roots of being, affecting every man and woman, body and soul.

The novel takes cognizance of three stages of the struggle of evolutionary theory with religion. Denunciation of the scientific doctrine characterizes the first stage. Since religion is truth, the syllogism read, and Darwinism is contrary to religion, Darwinism is not truth—and must be anathematized. When the mounting weight of evidence tipped the scales and proved the intrinsic truth of evolution, men were ready to discard religion as worthless, truthless. The syllogism now cut the other way. Despair through loss of faith characterizes the novel in this, the second stage. Having touched the two extremes in a wide sweeping arc, the pendulum now began to oscillate gently between, and the evolution-theology crisis entered its third and final stage. Reconciliation of the truths of religion with the truths of the Darwinian theory marks this final phase, the denouement and curtain of this significant Victorian drama.

Darwinism's invidious relation to theology was by no means the only side of the new theory which invited criticism. In the bickering and wrangling of the scientists over petty details, in the electric clash of highly charged personalities at scientific assemblies where more heat than light was produced, writers saw opportunity for rare jesting. Thus, the author of *Lord Floysham* could remark through one of his characters:

> Science is inherently progressive in consequence of the scientists rarely, if ever, agreeing with one another. What one advances, another contradicts, and this leads to the survival of the fittest.[1]

The earliest attacks on evolution were, for the most part, directed against evolution interpreted as the general theory of development of species, as well as against the theory of descent, allegedly from the apes. The denunciation of evolution and natural selection, predicated chiefly on a belief in the infallibility of religion and the Bible, is carried on with much vituperation and little logic. For the most part the novelists insist dogmatically that the theory of evolution is damnable because it is mechanistic, materialistic, or atheistic; that man's soul cannot be argued out of him by all the evidence of natural history. The more splenetic resort to invective and *ad hominem* arguments, tarring all evolutionists as "materialists" and "infidels" and the evolutionary hypothesis as the "monkey and molecule" theory.

Outstanding champion of the old order, Benjamin Disraeli, Earl of Beaconsfield, placed himself in the van of those ready to march in crusade against the new heresy. This was no new task for him. Twice before he had shown himself the uncompromising foe of evolutionary doctrine. In 1847, in his novel *Tancred,*[2] Disraeli trod violently on what he considered the heresy of Robert Chambers' evolutionary philosophy, a philosophy which he called "The Revelations of Chaos." In 1864, at the Oxford conference, with the General Election of 1865 in the offing, in a rock-bound conservative speech he took his stand with the angels; that is, with the side of the Bishops who did not want the common faith disturbed by any "gorilla" hypothesis. In 1870, Disraeli once more sounded off on the evolutionary theme; but in keeping with the new seriousness of the scientific issue, he discarded irony for polemics.

In order, apparently, to make people understand how entirely in earnest he was in his argument in the Oxford speech, Disraeli reproduced it in a careful and deliberate form in the General Preface to his *Novels,* published in 1870. In the Preface he lays his finger on the spiritual sore of the age: the denial of an active providence, the ascendance of Materialism.[3] For this, natural selection and the Higher Criticism are at fault. Masquerading as Progress, asserts Disraeli, the Darwinian hypothesis has led untutored minds to believe that the age of primeval inspiration is over, the era of truth now in; that since the one is incompatible with the other, religion must go. But, Disraeli avers, the "new" discoveries discover nothing new. "Even the most modish schemes of the day on the origin of things, which captivate as much by their novelty as their truth . . . will be found mainly to rest on the atom of Epicurus and the monad of Thales."[4] Nevertheless, whatever man may be biologically and physiologically, he held, man is something more; and this something more, which is over and above the physical and the chemical, is the most distinctive and essential thing about him.[5] It is the essence which man has in common with the angels.

In *Lothair* (1870) Disraeli returned again and again to the mystery of life as science uncovers and religion explains it. The story treats of that period of the late nineteenth century during which the Roman Catholic Church attempted to assimilate the church of England and combat Garibaldi's "Secret Societies of Atheism." Lothair, the wealthy and well-born hero, early becomes an orphan. His guardians are two: one, a Scotch earl, Presbyterian and a Whig; the other, Monsignore, a clergyman who seceded from the Anglican Church, entered the Church of Rome, and has risen to the rank of cardinal. The rival churches—pull devil, pull baker—struggle for the impressionable youth. After many spiritual slidings and backslidings he ricochets back to the Anglicanism into which he was born.

Through all the wrenches, tugs, twists, and jerks to which he is subjected at the hands of Catholics, Anglicans, atheists and agnostics, Lothair never loses his sense of the divinity of man. He is frankly disgusted when Monsignore describes for him the materialistic "religion of science":

> Instead of Adam, our ancestry is traced to the most grotesque of creatures; thought is phosphorus, the soul complex nerves, and our moral sense a secretion of sugar.[6]

From the mouth of a Syrian gentleman whom Lothair meets on the Mount of Olives pour the arguments which Disraeli included in his General Preface, against the naturalistic explanation of life. "There must be design,"[7] the Syrian assures Lothair. Previously, musing on the question of man's origin and destiny, Lothair reached the same conclusion. True, all the distinguished men of science with their arguments of "chance, necessity, atomic theories, nebular hypotheses, development, evolution, the origin of worlds, human ancestry" deny design. Nevertheless, they account for everything except the only point on which man requires revelation. They cannot, for example, explain the conscious element in man. The natural forces to which all creation is attributed are unconscious, and, Lothair reasons, the conscious cannot be derived from the unconscious.

The Church, however, explains what no one else pretends to be able to explain: Man is divine.[8]

Where Disraeli attacked the claims of science to a monopoly of the truth, Charles Reade snapped caustically in his novel *Put Yourself in His Place* (1870)[9] at science's presumption in questioning the truths of religion, and at claiming for itself any truth whatsoever. "There never was so infidel an age," his character Sir Guy Raby remarks. Sir Guy goes on to indict the sceptics who dare to question religion's evidences yet have nothing but naive credulity for the evidences of evolution—"the impudent lies and monstrous arithmetic of geology, which babbles about a million years, a period actually beyond the comprehension of the human intellect."[10] Through the mouth of Raby, Reade proceeds to accuse of chicanery the palaeontologists who amass evidence in proof of evolution, suggesting that the scientist "takes up a jaw bone that some sly navvy has transplanted overnight from the churchyard into the Lord knows what stratum, fees the navvy, gloats over the bone, and knocks the Bible down with it."[11]

Laurence Sterne used to call Romanism his bread and cheese. Whenever he was at a loss for a sermon he used to attack the Pope. With the coming of evolution Darwin and Huxley became the Pope for many Victorian novelists, and baiting the evolutionary atheist the new literary sport. In the novel *Margaret* (1872)[12] by Catherine Liddell (C. C. Fraser-Tytler) a Mr. Cargrave is the atheistic disciple of Huxley who is worried in the orthodox way for the amusement of a fashionable dinner-party. He is introduced to us, first, as the Atheist incarnate, with stock features:

> Anyone who has knowledge of the human countenance, or its diverse expressions, must know the peculiar cast of an atheists's face—the brow devoid of all grandeur, the fixed and soulless expression of the eye, the look of stern calculation, the intense materialism of his creed, pervading all.[13]

At the dinner party in spite of his "look of stern calculation" he and his materialism are overthrown by a young girl who later confesses that she is totally ignorant of scientific matters. She badgers Mr. Cargrave by daring him to explain how evolutionists can make their belief in purely natural causes agree with such a thesis as that of "spontaneous generation" that "we arise from a germ or germs in the waters of the globe." Evolutionists made no pretence at being able to explain the beginning of life, but the young Margaret Marshall finds no difficulty in rendering this one speechless with a crushing argument about the First Cause. She preserves the citadel of her faith and staves off a "lecture on protoplasm, isn't that the word?"

Like Charles Reade, his colleague in the school of sensational novelists, Wilkie Collins glances off from his main theme to express a gratuitous denunciation of those scientists who "consider the consolations of religion in the light of harmless playthings." Invited to march with the age, old Benjamin in *The Law and the Lady* (1875)[14] flays his generation with bitter sarcasm.

> Don't let's do things by halves. Let's go and get crammed with ready-made science at a lecture—let's hear the last new professor, the man who has been behind the scenes at Creation, and knows to a T how the world was made, and low long it took to make it. . . . Oh the new ideas! The new ideas!—what consoling, elevating, beautiful discoveries have been made by the new ideas. We were all monkeys before we were men, and molecules before we were monkeys. And does it matter? And what does anything matter to anybody?[15]

With that Benjamin dashes into the library to read *Puss in Boots* and *Jack and the Beanstalk,* and anything else he can find that doesn't march with the age.

Some years later Collins published another novel sneering at the new scientific ideas, his *Heart and Science* (1883). In the character of Mrs. Galilee, the vivisectionist, a satirical thrust is driven home against the female dilettante in evolutionary science. Mrs. Galilee, whose scientific studies cover "the fossilized indigestions of extinct reptiles," consummates her scientific ambitions with an "At Home to Science." The soiree is attended by three of those "superhuman men" Collins had excoriated in *The Law and the Lady* who "had each had a peep behind the veil of creation and discovered the mystery of life."[16]

Richard Doddridge Blackmore added his voice to those baying at the moon in the preface to *The Remarkable History of Tommy Upmore* (1884). To the author of *Lorna Doone,* ardently devoted to the Church of England, science is "the popular name for any speculation" of bold proportions, and evolution, a theory ruthless in its iconoclasm:

> What power shall resist the wild valor of the man who proves that his mind is a tadpole's spawn, and then claims for that mind supreme domination and inborn omniscience? Before his acephalous rush, down go piled wisdom of ages and pinnacled faith, cloud-capped heights of immortal hope, and even the mansions everlasting, kept for those who live for them.[17]

The novelty and notoriety of evolution and Darwinism attracted the attention of a great number of facile minds, ever ready to follow the newest fashion in art or science, in social or religious life. Having acquired a nodding acquaintance with some of the most publicized points of the new theories, these dilettanti constituted themselves a sect of pseudo-evolutionists or more frequently, critics without portfolio. That "we were once tadpoles, you know"; that men are descended from monkeys; and that "moths and butterflies flirt with each other as we do" were propositions requiring no great strength of intellect to grasp or to expound in a lively conversational way. Feeding this glibness was a commissary of books which George Gissing referred to as "specialism popularised," a writing "which addresses itself to educated but not strictly studious persons, and which forms the reservoir of conversation for society above the sphere of turf and west-endism."* Consequently, a colloquial acquaintance with these advanced theories, having found its way into certain sections of London society, soon likewise found its way into the Victorian novel. Fluent conversational evolutionists in the novel flung their wittiest darts at the newest scientific target, their zest in the game for the most part unembarrassed by definite knowledge.

The purported "monkey origin" of man lent itself particularly to shallow gibes and taunts. The fact that this aspect of evolutionary theory ran counter to what had been accepted as divine truth provoked also a number of attacks of a rancorous kind. Neither shallow nor rancorous, however, is the early piece of satire on man's descent penned by Charles Kingsley. The hair splitting tactics of Thomas Henry Huxley and his opponent Professor Owen, protagonists in the classic *Hippocampus minor* controversy, are the subject of Kingsley's chaffing, in the *Water Babies* (1863).[18] In this debate over the classification of mammals according to their brains, Owen had elevated man above the rest of nature on account of his brain; Huxley insisted that the mind as well as the body was a product of evolution, and pointed to his discoveries, such as that the *Hippocampus minor* is found in the brains of both man and ape, to substantiate this.

---

* Gissing goes on to illustrate this appeal to dilettantism. Just such a smattering of "culture" does the heroine of *New Grub Street* receive after parting with her husband, so that "though she had never opened one of Darwin's books her knowledge of his main theories and illustrations was respectable."

George Gissing, *New Grub Street* (New York, n.d.), p. 384.

Another instance is that of Nancy Lord in Gissing's *In the Year of Jubilee* (1894). Nancy dabbles in all the sciences with a view to impressing the young men of her acquaintance. She pretends the exact sciences are her favorite study; but inwardly she is thinking:

Evolution! She already knew all about Darwinism, all she needed to know. If necessary she could talk about it—Oh, with an air. But who wanted to talk about such things? After all, only priggish people,—the kind of people who lived at Champion Hill or idiots . . . .

George Gissing, *In the Year of Jubilee* (London, 1894), p. 172.

At its height, the Huxley-Owen row had been one of the sensations of the London "season." The catchphrases of the controversy had been on everybody's lips. *Punch,* for May 15, 1861, under the caricature of a gorilla bearing the sign, "Am I am Man and a Brother?" had published its own libretto to the passage of arms. It was dated "Zoölogical Gardens" and was signed "Gorilla":

Am I satyr or man?
Pray tell me who can,
And settle my place in the scale.
A man in ape's shape,
An anthropoid ape,
Or a monkey deprived of his tail?

The Vestiges taught
That all came from naught,
By "development," so-called, "progressive";
That insects and worms
Assume higher forms
By modification excessive.

Then Darwin set forth
In a book of much worth,
The importance of "Nature's selection";
How the struggle for life
Is a laudable strife,
And results in "specific distinction."

. . . .

Then Huxley and Owen,
With rivalry glowing,
With pen and ink rush to the scratch;
'Tis Brain versus Brain
Till one of them's slain;
By Jove! It will be a good match!

Says Owen, you can see
The brain of Chimpanzee
Is always exceedingly small,
With the hindermost "horn"
Of extremity shorn,
And no "Hippocampus" at all.

. . . .

Next Huxley replies,
That Owen he lies,
And garbles his Latin quotation;
That his facts are not new,
His mistakes not a few,
Detrimental to his reputation.

"To twice slay the slain,"
By dint of the brain,
(Thus Huxley concludes his review)
Is but labor in vain,
Unproductive of gain,
And so I shall bid you "Adieu!"[19]

Kingsley was present at the famous tournament between the professors on the *Hippocampus* question.[20] It moved him, as it did the writer of the *Punch* poem, to satire. At the time it called forth from him a little squib which circulated among his friends. In the *Water Babies* he repeated the jest in a mock-serious parody of the learned argument. His irony Kingsley directs first at Professor Ptthmllnsprts whom for the moment he identifies with Huxley. This professor is painted as a materialist who holds that nothing is true but what he can see, hear, taste, or handle. One of his strange theories he dared bring before the British Association: "that apes have hippopotamus majors in their brains" just as men have. The great fuss and fanfare over a minute question, and the ridiculousness of basing the whole concept of evolution on so trivial a point, irked Kingsley no little:

You may think that there are other more important differences between you and an ape, such as being able to speak, and make machines, and know right from wrong, and say your prayers, and other little matters of that kind: but that is just a child's fancy, my dear. Nothing is to be depended on but the great hippopotamus test. If you have a hippopotamus in your brain, you are no ape, though you had four hands, no feet, and were more apish than the ape of all aperies. But if a hippopotamus major is ever discovered in one single ape's brain, nothing will ever save your great-great-great-great-great-great-great-great-great-great-great-greater-greatest-grandmother from having been an ape, too.[21]

And then the author turns his irony on the anti-evolutionists with the observation that if an actual "hippopotamus" was discovered in an ape's brain, after the opponents of the descent theory got through identifying and classifying it, "it would not be one, you know, but something else." But *hippocampus* or no, Kingsley felt that Huxley had proved nothing. The ape may have tools very nearly as good as a man's but without a soul, God's directing force, he cannot use those tools.

In a clumsy *jeu d'esprit, Lord Bantam* (1872)[22] by Edward Jenkins, the *Hippocampus minor* and T. H. Huxley were once again the butt of the jest. Huxley, whose combativeness forced him into the public eye, was almost as frequently the subject of caricature and reviling in the novel as the descent theory he advocated. In Jenkins's satire the addle-headed Lord Bantam presides at a meeting of the Society for Developing the Mental and Moral Stamina of Women. He tells the society that according to the doctrines of Darwin and Comte distinctions between man and woman should be discarded since "there is ground to believe that the original of us all, in the simious shape" was a hermaphrodite.

Together with his wife, Lord Bantam attends the Grand Eclectic Symposium on "the highest evolution of the spiritual element, the physical basis, positive science." There Lady Bantam delivers a lecture on "Protoplastic Chaos" and a Professor Foxley on "The Hippocampus Minor and its relation to the Mosaic Cosmogony." The guests sit down to a supper where the menu is stocked with appropriate dishes such as *Rosbif protoplastique* and *Filets des enfans humains, au selection naturel.*[23]

On a serious plane and with less pronounced an animus toward Darwinian doctrine is a work of Edward Robert

Bulwer Lytton's which assails the theory of descent. For more than ten years following the publication of Darwin's *Origin of Species* in 1859, Bulwer Lytton showed positively no evidence of interest in Darwinism. In 1871, however, Lytton published *The Coming Race,* a Utopian treatment of Darwinism. In 1873 appeared *Kenelm Chillingly,*[24] a study of a Victorian Hamlet. The eponymous hero lards his vocabulary with choice Darwinisms, but the author has also made him the mouthpiece for a round disavowal of man's relation to the beasts.

On his return from college, young Kenelm makes a Coming of Age speech before the assembled tenants and guests of his father. He reminds his gaping, rural audience that man represents merely a stage in the course of evolution,

> . . . a development of some hideous hairy animal such as the gorilla; and the ancestral gorilla itself had its own aboriginal forefather in a small marine animal shaped like a bottleneck;[25]

and that some day the human race will be exterminated by a superior species. These ideas embodying the theme of Darwin's recently published *Descent of Man* (1871) and Lytton's own *Coming Race,* Kenelm's father, Sir Peter, denounces; but his audacious scion insists that they represent the notions most in vogue in London.

Well-equipped with these ideas, Kenelm draws upon them now and again in illustration of his homilies. He bitterly assails political and economic systems which imitate Natural Law in their disregard of the individual: "By Natural Laws creatures prey on each other, and big fishes eat little fishes"[26]—without doubt an echo of Darwin. And again showing his acquaintance with the *Descent of Man and Selection in Relation to Sex,* Kenelm remarks: "According to the Darwinian doctrine of selection fine plumage goes far in deciding the preference of Jenny Wren";[27] or later when he speaks of "the great gorillas who scratched off their hairy coverings to please the eyes of the young ladies of their species."[28]

But Kenelm's (read Lytton's) interest in the theory of man's descent did not preclude his attempting to score a palpable but not ill-natured hit at Darwin and the pretended omniscience of the evolutionists. In the absence of any explicit evidence on the manners of "our semi-human progenitors in the transition state between a marine animal and a gorilla," Kenelm insists that the so-called "cautious man of science" (Darwin) who proposed such a theory of man's origin is nothing but a creator of poetic romance. In the works of Scott, Cervantes, and Shakespeare rebuilding the past, yet claiming rank only as romances, Kenelm was prepared to find more authentic material than in the fantasy Darwin, "immeasurably more inventive," had written. Could the scientist, perhaps, cite one single chronicler who saw with his own eyes the gradual metamorphosis of one animal into another?

> I cannot conceive that even that unrivalled romance-writer can so bewitch our understandings as to make us believe, that . . . when some lofty orator, a Pitt or a Gladstone, rebuts with a polished smile which reveals his canine teeth, the rude assault of an opponent, he betrays his descent from a 'semi-human progenitor' who was accustomed to snap at his enemy.[29]

As one who dedicated his novel *Transmigration* (1874)[30] to "the preservation of our religion and our loyalty to our Queen," Mortimer Collins could be expected to join readily with the Darwin-baiters. His language differs little from that of his predecessors in this regard. He condemns his age and the shallow thoughtless people who regard as truth and view with admiration the "miserable misrepresentations of nature" of those who argue that we are developments of the lower creatures. "The Divinity is an ape, and behold! I am his prophet," is the cry.[31] No more than it could fool Bulwer Lytton, the "sham science" of these sophists who would affirm the non-existence of God and the monkey-ancestry of man cannot hoodwink Mr. Collins. "The thing cannot be done. Even a Darwin could not be grown from the chimpanzee . . . and there could be no easier form of the experiment."[32] From vituperation he turns to evangelical fervor, pleading with men to turn their interest away from our relation to the beasts that perish, to the higher nature of man, his relation to God.

For the malicious wit of Collins, Julia Kavanagh, the author of *John Dorrien* (1875),[33] substitutes mock solicitude to express her contempt of the notion of a bestial ancestry for man. In ridicule of the evolutionary doctrines of the atheist Oliver Black, a Mrs. Reginald remarks:

> I once knew a Mr. Poole who would have it that his first great-grandfather had been a monkey. It was evilently a comfort to him to think so. Well, I did wish he had a tail, for his sake; he would have liked to wag it, poor fellow! I believe they have given up the monkeys now. I dare say little Mr. Black thinks he came from an Ascidian jelly-bag.[34]

Most of the satires against Darwin's version of man's descent had been published by 1880. Beyond that date there appeared a few twilight pieces, more temperate than the earlier writings on the subject. Oblique and subdued is the satire of man's descent by Richard D. Blackmore in *The Remarkable History of Tommy Upmore* (1884). The novel was written, Blackmore tells us, to show that the body is beyond the comprehension of the mind. We know little enough about our body today, he points out; how can we know all about it for millions of years before either mind or body existed? How can we trace their joint lineage up to a thing that had neither a head nor a body.[35]

Tommy Upmore, the hero, is a natural phenomenon. Son of a fat father and obese mother, he puts to rout all the laws of heredity and the seers of evolution by showing a birdlike lightness; with the proper breeze, he actually can fly. His case throws consternation and dismay into the ranks of the "scientists" (Blackmore abhors the word, and pillories it between inverted commas), four of whom probe poor Tommy's peculiarity, each according to his own theory. Dr. Chocolous hopes to restore to the human race its primordial tail by avoiding sitting; Professor Brachipod offers a theory of "organic levigation" to explain Tommy and calls upon him "to revert to the age of unbroken continuity when that which is now called Tommy was an atom of protobioplasm."[36] Only Professor Megalow reserves judgment for further study, though long research into skeletons and fossils has quite convinced this new Darwin that man descended from lizards.

Marie Corelli, no follower of the Church in spiritual matters, in her continued resentment toward evolution refused also to follow the Church's lead after 1880 in forgiving and forgetting. The animosity she was to continue to express to the theory in most of her works is directed against Darwin's version of the descent of man in *Ardath* (1889).[37] The cynical observation of a converted atheist, decrying the irreligion of the aristocratic class to which she belongs, echoes Miss Corelli's private convictions.

> You must not forget that we have read Darwin,—and we are so charmed with our monkey ancestors, that we are doing our best to imitate them in every possible way,—in the hope, that, with time and patience, we may resolve ourselves back to the original species.[38]

Into the mouth of another converted atheist, the authors of *Paul Nugent, Materialist* (1890)[39] put a sneering disparagement of man's relation to the anthropoids. The first sign that the materialist Sir Paul's belief in evolutionary science has begun to slip is his jest at its expense:

> Quite time for another theory to crop up. Somebody for instance will discover that, instead of descending from apes, we go back to apes when we die.[40]

On a par with the "monkey theory" as a subject for jest in the novel was the popular conception of what constituted the "missing link" in the chain of evolution. Huxley's popularization of the "missing link" was perhaps responsible for the active interest in the idea, as evidenced by scattered reference to the term in a score of novels. W. H. Mallock's *The New Paul and Virginia* (1878),[41] a burlesque in the manner of Voltaire's *Candide,* caricatured the search for the "link" and several other aspects of evolution while aiming squibs at Positivism. The same ironic method Mallock had used successfully in *The New Republic* he tried again in the new work, directing his spleen against current atheistic philosophies as embodied and concentrated in his hero, Professor Darnley, "an atheistic professor, such as Tyndall."[42]

Professor Darnley is the author of three volumes on the *Origin of Life* which have cost him seven years of research in the infusions of hay and cheese. To escape from a nagging wife, the professor undertakes a cruise to Australia. On board he relieves the tedium of the journey by delivering a lay sermon to the passengers in which, spicing his remarks with quotations from Huxley and Tyndall, he shows them how "all existence is a chain, with a gas at one end and no one knows what at the other."[43]

Wrecked on the shores of a desert isle alone with the beautiful Virginia St. John, wife of an Anglican bishop, Darnley attempts to convert her to the new philosophic religion—Positivism. The sudden appearance of a hideous, hairy apparition at the window, however, distracts the professor from his proselytizing and awakens the evolutionist in him. It is the "missing link," he is certain. He must hunt for it. Virginia is frankly sceptical: "as to thinking that we really came from monkeys of course that is too absurd."

The professor is nettled.

"Do you know," he snorts, "that till I have caught the missing link the cause of glorious truth will suffer grievously? The missing link is the token of the solemn fact of our origin from inorganic matter. . . . It is through him that we are related to the stars."[44]

Patiently the Professor tracks the beast, finds its lair, collects in a card-paper box a large number of its parasites which he subjects to microscopic analysis.

The "missing-link" turns out to be a tame monkey.

For those who might have doubted the genuineness of his parody, the author appended, in the manner of Thomas Love Peacock, the context of the quotations which the novel exploits, from Professors Tyndall, Huxley, Clifford, from Harriet Martineau and Frederick Harrison.

A Professor Marvin (Darwin?) is represented in F. G. Walpole's *Lord Floysham* (1886) as engaged, like Professor Darnley, in the search for the missing link. His presence at a dinner party in the company of women in low-cut gowns suggests to one of the characters the possibility that the Professor has carried his search for new physiological developments into the drawing room. But Captain Luttrell, a Life Guardsman with a contempt for "Marvin's theory" of his kinship with the apes, gives warning that he will punch Marvin's head should the Professor dare to suggest the presence of a trace of the monkey in the beautiful Miss Thornton. She, he gallantly insists, echoing Disraeli and current jokes, is descended rather from the angels. The Captain avers that the queer cranial formations of the scientists themselves which he had leisure to observe at a lecture on evolution must have suggested the theory to them. Though Lord Floysham, the hero, believes Captain Luttrell too severe upon the men of science, he himself does not hesitate in describing the growth of scepticism to sneer at "Nothingism with Evolution as its prophet."[45]

In John Davidson's *Earl Lavender* (1895)[46] a burlesque representation of the Missing Link, in the manner of W. H. Mallock, is held up for the reader's derision. In the course of their attempted escape from wife and sweetheart, the two heroes of the novel flee precipitately in a stolen hansom cab. Not until they have reached a distant grove do the errant males alight, only to be faced by a startling spectacle. Before them is what appears to be a compound monster, the upper half an ape in Scotch bonnet, blowing a bagpipe; the lower half, human, attired in tartan. Earl Lavender immediately concludes that this is the missing link, "a bi-formed beast, one portion of which became the progenitor of the human race, the other of the monkey tribe."[47] But when the human half cries out, "Save me!" Lavender realizes that the "link" is an ape sitting on a Scotsman. The ape—an orang-outang—is finally overcome, and Lavender rejoices in the survival of the fittest. They proceed to celebrate the funeral of "the bestial part of the Missing Link."

Such attacks on evolution as these, some spiced with good humor, others, wrathful and splenetic, it is patent, make no pretence at a refutation of the theory. Instead, they seem directed for the most part to stirring up the prejudices of unthinking readers. The roll call of the propagandists includes the names of most of the ranking fiction writers of the age: Disraeli, Charles Reade, Wilkie Collins, Blackmore, Bulwer Lytton, Corelli, Mallock, Davidson. So united a front of popular novelists at work on public opinion could not have been without its effect in conditioning the responses of the average Victorian reader. Since the difficulties which stood in the way of the acceptance of evolution in the '60's, '70's, and '80's were more doctrinal than scientific, the opinionative objections of the best-selling novelists counted heavily against the theory.

CHAPTER SIX

SATIRES ON DARWINISM

More comprehensive and intelligent by far than the outbursts against the theory of evolution and of man's descent were the fictional treatments which turned not on the doctrine of development but on Darwin's thesis explaining it. These compositions required and give evidence of an intimate knowledge of the doctrine of natural selection. Fortuitous modifications, sexual selection, the struggle for existence and survival of the fittest are all gibbeted by these literary hangmen. The point of the satires is by no means obscure, though occasionally overlaid with erudition. Thus, though George Meredith refused Samuel Butler's *Erewhon* (1877) for Chapman and Hall on the ground that it appeared to be a philosophical work and Trübner and Company returned the manuscript without reading it, thinking it to be a treatise on contagious diseases,[1] when the book was ultimately published it proved extremely popular. The public at least could tell a hawk from a handsaw.

Samuel Butler wrote three novels. Two of these, it cannot be gainsaid, owe their direct inspiration to Darwin; the third is merely a sequel to the first. Prior to the writing of these novels, Butler composed and published a number of semi- and pseudo-scientific treatises of which the *Origin of Species* was progenitor.

The *Origin of Species* was published in the autumn of 1859, and Butler arrived in New Zealand about the same time and read the book soon afterwards. In 1880 he wrote in *Unconscious Memory:* "As a member of the general public, at that time residing eighteen miles from the nearest human habitation, and three days' journey on horseback from a bookseller's shop, I became one of Mr. Darwin's many enthusiastic admirers, and wrote a philosophic dialogue . . . upon the *Origin of Species*."[2]

Soon after the original appearance of Butler's dialogue, a copy of it fell into the hands of Charles Darwin, possibly sent to him by a friend in New Zealand (possibly, too, by Samuel Butler himself). Darwin was sufficiently struck by it to forward it to the editor of some magazine, with a letter (1863) characterizing Butler's effusion as "remarkable from its spirit and from giving so clear and accurate a view of Mr. D's theory."[3]

The Dialogue and subsequent correspondence are interesting as showing that Butler was among the earliest to study closely the *Origin of Species,* and also as showing the state of his mind before he began to think for himself. His work consisted merely of a restatement in concentrated form of the main tenets of Darwin's thesis. Of original thought there is not a trace; but it does give evidence of what Darwin called Butler's "rare powers of writing" and of the preparation of that soil in which the seeds of *Erewhon* and all other works was to grow.

The cloud of ill-will and hostility which in later years was to fall between Darwin and Butler now was a babbling sunny brook. In these first years after his return to England in 1864, Butler sought anxiously in correspondence for the great evolutionist's approval and purred appreciatively when he got it.[4] He is the adoring shepherd himself in a reply to a query by Darwin (September 30, 1865) as to his future plans. "I always delighted in your *Origin of Species* as soon as I saw it out in New Zealand—not as knowing anything whatsoever of natural history, but it enters into so many deeply interesting questions, or rather it *suggests* so many, that it thoroughly fascinated me."[5]

Erewhon was Butler's first novel. Published in 1872, thirteen years after the appearance of Darwin's *opus magnum,* the work contains an ingenious extension of the evolutionary theory suggested by the *Origin of Species*—

the section of *Erewhon* entitled "The Book of the Machines." The three chapters of the novel which comprise this section not only embody Butler's most fertile adaptation of the doctrine of evolution, but likewise contain in embryo the criticism of orthodox Darwinism which he was later to formulate in *Life and Habit,* in *Evolution Old and New,* in *Luck or Cunning?,* and in *Unconscious Memory.* The chapters on the Machines in *Erewhon* were based on two short pieces, "Darwin Among the Machines" (1863), and "Lucubratio Ebria" (Drunken Meditation) (1865); and on another piece, "The Mechanical Creation" (1865), published shortly after his return to England.[6]

"The Book of the Machines" is a significant part of the story of *Erewhon,* motivated by the episode of the watch which occurs quite early in Higg's adventures in the topsy-turvy land. On Higg's person, when he is brought before the chief magistrate, is discovered a watch, the possession of which, though of no significance to us, was to the Erewhonians a capital crime. Higgs knew nothing of this. When, however, he was brought to the town's museum and introduced to a room whose show cases were filled with broken parts of highly developed machinery, the realization dawned upon him that anything mechanical was contraband.

Five hundred years before, machines, all but the simplest, had been destroyed after a violent revolution. It was an outbreak against the mechanical such as Ruskin would have applauded with both hands, throwing his last thousand into the war chest. Strangely, a philosophical manifesto coldly analyzing the possible future evolution of mechanical apparatus was responsible for inciting the destructive frenzy against the machine. Before leaving the country Higgs was able to get possession of a reprint of the treatise which had caused the havoc.

The theory of evolution informs every page of this document. It begins by calling attention to the fact that consciousness as we know it, once did not exist. At one time fire, at another rock, at still another, plant life was the last word in consciousness. Yet with this evidence of change before us we rest assured that animal, human life is the highest development possible. Is it not possible that another and higher form of consciousness than the one we know may come into existence? Are there any indications that a higher form is likely to supersede human development? Yes, answers this ancient philosopher of Erewhon, machines show unmistakable signs of superseding man. True, they have little consciousness as yet, but remember that the mollusc had little at one time. Moreover, a great deal of machine consciousness that we are given to call "mechanical" in no way differs from mechanical reactions in plant and animal which we call "conscious"; as the action of a fly-eating plant, the growth of a boy or a flower.

Natural selection has played its part among machines as in the animal and vegetable kingdoms. Witness the advantage of any current species of mechanism over an earlier counterpart; witness the rudimentary organs, useless now, yet serving to mark descent from some ancestral type; witness (as in the watch) the diminution in size of machines which has attended their progress and development, paralleling that from the dinosaurus and plesiosaurus and other gigantic reptiles. Machine improvement still continues. Daily we add to their beauty and delicacy. Daily we give them power not only to do our bidding, but to regulate themselves. To what end? To the creation and grooming of our successors, answers the writer.* We are committing slow suicide. In the course of

---

* Butler, in his *Notebooks,* accused George Eliot of plagiarizing her chapter on machines, "Shadows of the Coming Race," in *Impressions of Theophrastus Such* (1879) from his *Erewhon.* He was highly flattered: "In one year I made Mr. Darwin write a book ("Erasmus Darwin") and George Eliot crib a chapter."

Samuel Butler, *Further Extracts from the Notebooks of Samuel Butler,* ed. by A. T. Bartholomew (London, 1934), p. 90.

ages we shall find ourselves the inferior race; inferior in power, in self-control, in freedom from evil passions and turbulent emotion. As slaves to the machines we shall minister to their wants, physic their ills—*in perpetuam:* "for what machine dies entirely and in every part at one and the same instant" like a one-horse shay?

Note the extent to which men are already the slaves of machines. If the latter were all destroyed, the human race would die in six weeks. And then consider what a short time, comparatively, it has taken machines to gain this ascendancy. Animal and plant life has been in this world for, say, twenty million years; machines, as we know them, were born relatively about five minutes ago.

But, the objectors say, the machines require man to tend them, to feed them. True, comes the answer, but whereas a generation ago plows required horses and men to give them food, that is, energy, now we see them feeding themselves. But at least machines can never reproduce themselves. Perhaps not in the ordinary sense; and yet we see machines making other machines. Yes, but still man is always the agent, the director in such activities. Very well, answers the philosopher, wherein does man's role in this respect differ from that of the bees which act as intermediaries between certain kinds of flowers and are absolutely necessary to their reproduction? Machines had no stomachs; they have them now; the machines have no complete system of reproduction now; in the course of time they may. The germ is there; the complete system may not be far off. Though it differ from the human, nothing demands that the machine follow the human method any more than the human or animal follows that of the plant. Each part of the engine, we may anticipate, will be bred by its own special breeders whose function is to breed that part and that part only. The combination of parts may well be the function of another part of the mechanical reproductive system.

Though some may point to advantages accruing from a life spent in subservience to the machines, the philosopher will not countenance the thought. It is too spiritually revolting.

Could I believe that ten hundred thousand years ago a single one of my ancestors was another kind of being to myself, I should lose all self-respect, and take no further pleasure or interest in life. I have the same feeling with regard to my descendants.[7]

Only one serious attempt was made to answer the manifesto and that was as a voice crying in the wilderness. Its author held that machines, far from being identities or animisms potent to triumph over mankind, are merely extra limbs aiding the rise of man. Every fresh invention, he pointed out, is an additional resource of the human body; a railway train, a "seven-leagued boot which five hundred may own at once." Highest in the scale of animal life are those millionaires who can command the full complement of limbs with which mankind could become incorporate.

But the Erewhonians were true Darwinians. Realizing what natural selection had been doing for the machines in the past, and what chance circumstance might do in the future, they vowed to put an end to the Frankenstein monsters. Death was decreed for the machines—all but the simplest; and death was the measure dealt out.

It was with something of a shock that Butler learned, on the publication of *Erewhon*, that some of its reviewers had thought he was poking fun at Darwin. To the young writer of thirty-seven even the suspicion of his having rollicked with the sacrosanct *Origin of Species* was unbearable. He ridicule a theory which he accepted without a reservation? He gibe at a man whose friendship he prized? At once, on May 11, 1872, he dashed off to Darwin a somewhat breathless letter to the effect that the subject of machines was "obviously an absurd theory without a particle of serious meaning, written simply to show how easy it is to be plausible and to defend an absurd proposition by a little ingenuity." He disclaimed all intentions of being disrespectful to the *Origin of Species,* a book for which he could never be sufficiently grateful, though well aware "how utterly incapable I am of forming an opinion on a scientific subject which is worth a moment's consideration."[8]

And in the preface to the second edition he added (in accordance with a promise made in the letter) that nothing could be more distasteful to him than any attempt to laugh at Mr. Darwin for whose work he had the profoundest admiration.[9] To all intents and purposes he had set himself quite straight "about having intended no villainy by the machines."[10]

Butler may have salved his own conscience by his effusive apology but what he may not have been doing consciously he was doing subconsciously. There is no doubt that at this early date he was dissatisfied with Darwinism in at least one regard: Darwin's imputing the origination of modifications to "fortuitous circumstance." It requires no microscope to discover in "The Book of Machines" an implied criticism of this phase of natural selection.

"Fortuitous modification" is the foundation-stone of Butler's machinery fantasy. Darwin's theory of evolution by the preservation of modifications created by "chance" had, it was said, "banished mind from the universe." If, then, the development of species can be premised as proceeding not with purpose or design but haphazardly, certainly the conception of machines' acquiring the proper modifications to become a species more highly developed than man is a tenable hypothesis. This was humor, but it was likewise sound reasoning.

Further, if by logic a machine could bear comparison part by part with man, the next question asks itself: If design is necessary to make a steam-engine can it be absent from the making of man? To Butler the answer demanded by this analogy was an emphatic, "No!" Evolution, whatever else it might be, was then a purposive, intellectual process. The "omnipresence of intellect" must be granted.

Not chance then but design, Butler held, governs organic development. Just how design operated in the evolution of species Butler likewise suggests in "The Book of the Machines." Machines are only supplementary limbs which man has made for himself through his intelligence and sense of need. If our sense of need has called into play an intelligence to the creation of extra limbs in the form of machines, could not a sense of need have operated also in the creation of adaptations or extra limbs for primordial organisms? But here again was involved a virtual denial of Darwinian "accident." Here again was Butler pressing the point of his sword into the very heart of what he later provokingly was to designate Charles-Darwinism. The point was this, "Luck or Cunning?" Did living species originate, vary, and flourish by the chance accidents of a Natural Selection without direction; or by virtue of a mind seizing its opportunities, as man, to make the best it could of matter, space, and time? In accepting this latter alternative, that an organism exercises cunning in an effort to adapt itself to its environment and to become more efficient, Butler aligned himself with Lamarck against Weissmann and other neo-Darwinians who went beyond Darwin in their espousal of natural selection not only as a true cause, but as practically the sole cause of descent with modification. In his succeeding books on evolution Butler came to look on Darwinism as a mere backsliding from the true faith long before delivered by Lamarck and Erasmus Darwin; and he frankly and definitely announced himself as champion of the teleological theory of evolution against the mechanistic principles of the ultra-Darwinians.

Butler's deadly logic carried the development of the machine from an acquired consciousness to an acquired will to dominate, then through to an imagined victory over man. The Erewhonians destroyed all machinery sufficiently complex to suggest consciousness in the process of emergence. Mankind, however, Butler implies, is still faced by this bane. Surely this growing mechanization of life is no less a danger to us than it was to the Erewhonians. That which the Erewhonians learned to dread we too must learn to fear: that with man's dependence upon machines as extra corporeal limbs man himself has ceased to evolve; that eventually the whole body may become purely rudimentary, "an intelligent but passionless principle of mechanical action." That we are on our way to degeneration Butler logically deduces. Physical and intellectual superiority have come to count for naught—a prelude to degeneration, surely Instead, men have learned to admire "those mighty organisms," our leading bankers and merchants, possessing a complete set of machine-limbs —or wealth. It is these millionaires men are ready to recognize as "higher than themselves in the scale of animal life." To these they pay deference as the fittest.

Following the disturbance over the machines, there was for a while a panicky stir among the Erewhonians over the rights of vegetables and animals. The Erewhonians had been forbidden to eat meat; and when it was discovered that both plants and animals have had a common ancestry, philosophers ruled that plants have consciousness, and like meat must be ruled out of the diet.

To establish this principle of plant consciousness, Butler drew upon his newly formed doctrine of "unconscious memory." This he was later to develop in thesis form in

Life and Habit (1877). "Unconscious memory" was Butler's particular contribution to Lamarckism. According to this theory, the acquired experience of the parent is passed on as an instinct to the embryo. The new organism is literally identical with its parents, and these in turn with theirs, and so back by an infinite regression to the hypothetical "first form" of life. Consequently, the germ, or impregnate ovum or embryo, is simply repeating a performance which it has gone through an immeasurable number of times before.* Hence its absolute familiarity with the circumstances, its serene procedure—always knowing precisely what to do next, and never at a loss.

The rose-seed does what it now does in the persons of its ancestors—to whom it has been so linked as to be able to remember what those ancestors did when they were placed as the rose-seed now is. Each stage of development brings back the recollection of the course taken in the preceding stage, and the development has been so often repeated, that all doubt—and with all doubt, all consciousness of action—is suspended.[11]

Driven by this logic to believe in the unconscious intelligence of plants as well as of animals, the Erewhonians first treated plants as taboo. Then, since they could not live without plants or animals, the pragmatic Utopians decided that sin is better than death and so overcame their squeamishness.

---

* Marie Corelli's faculty for romanticizing science was exercised on Butler's principle of inherited memory in her novel *Ziska* (1897). In that work, Doctor Dean carries the principle to matters of love, and seeks to justify it:

The doctrine of evolution proves it. Everything that we were once has its part in us now. Suppose, if you like, that we were originally no more than mere shells on shore,—some remnant of the nature of the shell must be in us at this moment. . . . I maintain that if it is decreed that the soul of a man and the soul of a woman must meet,—must rush together, not all the forces of the universe can hinder them. . . . For mark you—in some form or other they have rushed together before! Whether as flames in the air, or twining leaves on a tree, or flowers in a field . . .

Marie Corelli, *Ziska* (New York, 1898), pp. 190-192.

By his philosophic contribution to science, Butler was able to restore design to the universe, to reconcile teleology with evolution. His *a priori* principle of inherited memory, whether applied to plants or animals, suggested a plausible explanation of the similarity of child and parent, and of the difference. Variation, then, was due not to "luck" but to the striving (or "cunning") of the individual in adapting itself to its environment, and handed on by the inheritance of "unconscious memory" or "habit"—all part of a grand design working itself out consciously in the evolution of life. All the ideas about evolution and the Life Force which George Bernard Shaw learned to expound, and produced dramatically in his *Back to Methuselah,* were but a rephrasing of these evolutionary concepts.

Almost every aspect of Darwinism—and even Samuel Butler's neo-Lamarckism—are subjected to mockery in the all-embracing travesty on evolution, *That Very Mab* (1885),[12] by May Kendall. The novel is as much a fantasy as Butler's Utopian *Erewhon.*

Queen Mab who was driven out of England by the Puritans returns to her native land in the guise of a butterfly. Though she counted on the friendship of the bees, birds, and insects, she finds them indifferent to her, with egos inflated by the fact that scientists have composed monographs on them. The queen bee even volunteers to give the visitor an account of the formation of hexagonal cells by natural selection, as culled from the pages of the *Origin of Species.* Slowly it dawns on the fairy queen that she is an anachronism in England, that the fashion of believing in fairies has been snuffed out by scientists "who have put everything under their microscopes except stars and First Causes."[13]

Mab has the misfortune to be caught in the butterfly net of a scientific professor who puts her in a glass bottle and takes her home. The Professor's son Walter who has been brought up in strict agnosticism surprises his father by ecstasizing over the supposed butterfly.

"We have at last discovered the origin of religion," cries the Professor. "It isn't the Infinite. It is worshipping butterflies. . . . The boy has returned to it by an act of unconscious inherited memory, derived from Palaeolithic Man, who must, therefore, have been the native of a temperate-climate, where there were green lepidoptera. Oh, my friends, what a thing is inherited memory!"[14]

An argument ensues between the Professor and a clergyman "veering towards the modern school" over the proper interpretation of a butterfly's coloring. The Professor, a disciple of Darwin, insists on explaining the tint and plumage as striking examples of natural and sexual selection. The theologian, a disciple of Paley, will have none of this. The butterfly's coloring, he insists, is a proof of the intelligent construction of the universe, and the argument from design; and he urges the scientist to become a convert to the new scientific religion. But a poet in the company who calls upon them to desist draws upon himself the fire of the Professor who stigmatizes poetry as "retarding the great progress of Evolution."[15] In the midst of the altercation, Queen Mab escapes.

With a wise old Owl as guide and philosopher to initiate her into the changes which have taken place since her exile, the fairy queen goes on to London. The Owl, with an impartial contempt for everything and everybody, puts Queen Mab through a course in applied evolution, its principles and practice, in explaining the strange urban sights to her. The people and vehicles rushing up and down Cheapside he interprets airily as "competition, struggle for existence and all that." Policemen are products "of the higher civilization evolved to put the lower civilization into prisons." Prisons are a kind of hothouse for the culture of feeble moral principles that the struggle for existence has been too much for. That evolution should turn out moral principles that are not first class, the sage attempts to excuse: The moral principle, he avers (with perhaps a glance of acknowledgment at Herbert Spencer and at Professor Henry Drummond whose *Ascent of Man* popularized the law of altruism), is the latest product of evolution, and more time and practice is needed in turning it out.[16] Even the British manufacturer is given a rap on the knuckles. Of him the Owl tells Mab:

> He *will* get all he can for his money. In him you may see a typical, beautiful example of the Survival of the Fittest. He worked his way, by means of native moral superiority and pure chocolate composed of mortar and molasses tinted with sepia, right from the gallery into one of the very best reserved seats.[17]

Meanwhile the Professor has gone intellectually berserk. Revelling in the thought that evolution has not completed her work he foretells the doom of religion and democracy and envisages the Being of the Future. Bishops, he tells his friend the Bishop, will go the way of pterodactyls; for in the economy of nature "bishops are an unnecessary organ" which in the course of time will atrophy and degenerate out of existence. Democracies, he tells his friend the Democrat, will be represented in some geological record of the future by a conglomerate of "dust and bones, and ballot-boxes, and letters in the *Spectator*."[18] Our successor, the Being of the Future, the Professor predicts, as superior to us as we are to the monad, will not be anything of the human mould. He tells off on his fingers the four classes of vertebrates — fishes, reptiles, birds, and mammals.

> Only four! . . . Why it is preposterous, it is inconceivable that we should stop at four! . . . Where is the fifth! Cannot Natural Selection, Struggle for Existence, Variability, and Survival of the Fittest between them furnish a fifth class of vertebrates? I demand it in the name of Science and Evolution. We have been human long enough.[19]

But the Democrat will not be cowed by the Professor's fee faw fum. The ballot, he insists, is better than Natural Selection. "Natural Selection is all very well; but it does not know what we want. We do."[20]

Finally the Professor takes to wandering through churches gathering material for a book, "Competition, and the Survival of the Fittest, as displayed in Modern Sectarianism." His son Walter who, according to the Owl, has reverted to type, has learned to regard birds "not as products of evolution, but as things suitable to shy stones at, and to be treated with contempt and catapults."[21] Rather than face such dangers Queen Mab flies off to settle on one of the Admiralty Islands.

The chaffing of the evolutionary scientist who admits no principle except that of natural selection is carried on in Thomas Longueville's *Life of A Prig* (1886).[22] The hero of that novel on a theological odyssey has tasted all the religions of the earth, finally to arrive at agnosticism. He becomes tutor to a young philosopher, the Honorable William Heriot, who acknowledges no laws except "the preservation of the Ego, the greatest good of the greatest number, and the survival of the fittest."[23] In accordance with these principles the youth suggests upsetting all the laws of morality to the end of permitting a plurality of wives for the fittest men; for they are "the fittest husbands for the greatest number of the fittest wives."[24] Since he would debar the unfit from marrying at all, the young man is troubled by the question of who are the fittest. Should he debar those intellectually powerful and weak physically; or weak in mind and strong in body? At any rate, he reasons, the fittest once chosen, natural selection (as he conceives it) will begin to operate: the fittest will be allowed to select as many wives as they please.

But the logic of his theory almost overwhelms him. By these standards puny young Heriot himself is not fit to survive. Not only must he never marry, but, he decides, it may be his duty to destroy himself. The tutor humors him in this game of logic-chopping but when he turns on his instructor and suggests mating him with "matter—not mind—a large muscular woman with a simple untutored mind," the hero concludes it is high time he gave over his employment.

John Davidson's farcical treatment of the survival of the fittest doctrine in *Earl Lavender* (1895)[25] might seem paradoxical in the light of the poet-novelist's avowed materialistic atheism. But Davidson's atheism was more a rebellion against the strict religious training of his mother and preacher-father than the outcome of well-reasoned scientific or philosophic thought.[26] In his scientific ideas Davidson was primarily a poet and not a scientist or philosopher. Hence, he based his opposition to natural selection largely on mood: the thought of descent was repulsive to his imagination.[27]

Earl Lavender attacked not merely the principle of evolution but the numerous young men who came out every day with something they considered new and revolutionary; "fantastical creatures" of the *fin-de-siècle,* products of the pseudo-philosophy and foolish fiction of the day. As in his four other novels, Davidson relates in *Earl Lavender* the picturesque tale of a hero who follows out, in the Don Quixote manner, a particular pose or point of view with a mad-like persistency; a story which he decks out in hurly-burly wit, horse play, and absurd situation.

Sir Harry Emblem, a young man of twenty-five who has just run away from his bride of eight days, resolves to undertake an experiment he has long contemplated—to dedicate himself to Evolution. He "got Evolution" just as some people "get religion." Under the warming influence of his new-found gospel he has reached the conviction that he is the "very fittest human male at present breathing," and decides to live "the evolutionary life." As an apostle of evolution he will seek to convert others to the new way, and to find for himself the fittest woman with whom to mate. He enlists a Mr. Gurdon, a runaway from the brink of matrimony, under his banner.

Without a single penny, but confident in the power of Evolution to protect the fit, they set forth "to conquer the world for Evolution." In keeping with their new status the two evolutionists change their names: Mr. Gurdon becomes Lord Brumm, and Sir Harry becomes Earl de l'Avenir, corrupted to Earl Lavender. To disseminate their doctrine and challenge Evolution to a direct justification of the law of the survival of the fittest, the penniless Darwinians, having dined sumptuously, throw themselves rather arrogantly on the mercy of a group of journalists eating at the same restaurant. To these Earl Lavender announces his catchword: "The fit shall survive, and Earl Lavender is the fittest."[28] In return for having heard him usher in the first day of the Year One of the Evolutionary Era, the Guild of prosemen are granted the privilege of providing payment for the food the two have consumed.

Lord Brumm weakens in his resolution when Lavender complacently orders a cab to drive them to a cafe for an elaborate supper. But Lavender is confident.

> Nothing grieves Evolution more, my dear Brumm, than half-belief. We must not proceed on the assumption that it is easier for Evolution to provide the cost of a dish of macaroni than that of a supper of four courses . . . to be worried and economical ill becomes the apostles of that power which wasted countless ages in fashioning indolently one little world.[29]

Once more, leaving the cab waiting outside, they trust to Evolution to defray their expenses in the cafe. A beautiful veiled woman at a neighboring table attracts Lavender's attention. His mind is made up instantly. That woman, he decides, is the fittest mate for him. Brumm, however, who pretends to misogyny, disapproves on counter-evolutionary grounds. Woman being undoubtedly a most inferior creature, Brumm captiously suggests,

Evolution should have by this time provided some other means for the continuance of the race, and have allowed man, the fittest, to survive alone. But for Earl Lavender the mere survival of woman is sufficient attestation to her fitness.

Then to the amazement and amusement of the patrons, Lavender delivers a lecture, his second that evening, urging all present to join with him in challenging Evolution to vindicate their fitness. Bestow your wealth on the poor and begin a life of beggary, he invites. A waiter excitedly calls for the police, but before they can arrive the Veiled Lady whom Lavender has singled out as his intended mate comes to their aid, pays their bill, and hustles them out into their cab.

Accepting this as merely the fulfillment of the intention of Evolution, Lavender proposes to the Lady that she wed him as the fittest mate, but she derides his foolish fancy. She leads them to the secret meeting place of a Society of Flagellants, where, willy-nilly, the evolutionists join in the occult rites. But Lavender will not be gainsaid. He leaps to his feet and announces to the assembled victims of perversion: "Behold in me the purpose of the ages, Earl Lavender, the fittest of men."[30] Both he and Lord Brumm are thrust into the night.

Maud and Mrs. Scamler catch up with the two evolutionists, but Earl Lavender, now Harry Emblem again, will have only the fittest woman for a mate. He suggests finding her by "Natural Selection." He will invite delegations of women from every country, live with each member for a few days, and then by natural selection choose the fittest mate. Dissuaded from this by his wife, he accepts as a compromise measure a second visit to the Flagellants. There the neo-Darwinians transgress again and are haled before the Nameless One.

"Damn the theory of evolution," the Nameless One intones. . . . "It is time it were played out, like the hundred theories that have preceded it.* Had there been no monkeys this theory would have been a very subordinate theory indeed; it could never have made such headway against *two* missing links."[31]

This sophistical condemnation of his credo brings Harry back to his senses and back into the arms of his wife.

The bitterness that fouls most of the satires against evolution is absent from these fictional treatments of Darwinism. The mood is Horatian, sweet reasonableness with a dash of tartness. Not only is a thorough reading and an intelligent understanding of Darwinian doctrines shown by the authors, but also an awareness of the extension of the philosophy of evolution beyond the field of biology. May Kendall's *That Very Mab* sardonically represents the manufacturer inflated with the success gained by cheating, preening himself on his fitness to survive. Earl Lavender's farcical struggle for existence is an economic not a biologic one. Butler's cynical *Erewhon,* however, anticipated both of these in treating of evolution's application to economics. The Erewhonian "Book of Machines" prophesies satirically the profound truth that the development of machinery will not result in Utopia but in man's becoming a machine-tender at the expense of his human qualities. In successive romances which, like *Erewhon,* were to make pretence at prophecy according to evolutionary law, the question of whether man's state in the future will be nobler and happier through his use of machinery, became a dominant theme. The ground was shifted then, once the truth of evolution had been established, from a conflict over the truth of the theory to the larger issue of whether implicit in the course of evolution there lie the germs of future progress or the germs of decay.

* Cf. also Davidson's remarks in his essay, "Evolution in Literature": "Though evolution is bound to rule the minds of men for hundreds of years to come, intelligence knows it will be dismissed, as the idea of creation is being dismissed now."
John Davidson, *The Man Forbid and Other Essays* (Boston, 1910), p. 152.

CHAPTER SEVEN

The Loss of Faith Through Evolution and Darwinism

The Bible does not pretend to be a text book of natural science. But there were those in the '60's and '70's to whom its account of the six-day act of creation was literal truth, indubitable as the fact that the Bible was revelation. The whole superstructure of their faith was based on the truth of Scripture; and if that truth were impugned, inevitable was the destruction of belief. If the Bible lied, then God lied, for the Bible was the word of God, they had been told. And if God lied. . . . One theological authority had dared to put the theological argument squarely before the world: "If the Darwinian theory is true, Genesis is a lie, the whole framework of the book of life falls to pieces, and the revelation of God to man, as we Christians know it, is a delusion and a snare."[1] When the mounting weight of evidence day after day, week after week, year after year continued to bury beneath it all doubts of the validity of the hypothesis, the clergy had cause to regret having put the alternative so strongly. Darwin had proved his case. Evolutionism now stood forth as Truth, and dogmatic theology shrank back, given the lie.

Hardly had the theory of evolution and the Darwinian hypothesis begun to smash ruthlessly into the pinnacles of faith when the religious problem novel reappeared, after a Rip Van Winkle's sleep from the days of Newman's *Loss or Gain* (1848), and Froude's *Nemesis of Faith* (1849). By this time for Victorian fiction to paint life as it was seemed not at all strange. The novel had long ceased to be the diversion of an idle hour. It had grown more ambitious. It was the vehicle of almost all thought for which a large audience was desired. As such it had more of the interest of the essay and less of the interest of the story. Unlike the old days when it was not only heresy but bad taste for people in polite society to confront problems, the Victorian novelist now not only dared to point out social evils and represent the wrongs of classes, but was ready to take in new provinces, the great questions of whence and whither which haunted the abysses of thought. Fiction had come of age.

That conflict which is the essence of drama, novelists found in the clash of evolution and religion. Their treatment of the theme was sober and serious as befitted the subject and the mood of those earnest Victorians. For those who had pinned their faith on the ultimate loser in the conflict there was tragedy; for those who carried the ribbon of the stronger side, there was little satisfaction in the victory, for it was an unpopular one. Nevertheless, the sympathy of the authors themselves was for the most part with the science and against the creed.

The novel of Religious Doubt and Unfaith spread over a period of thirty years, but the incidence was greatest during the first twenty years from 1870 to 1890. Drawing heavily on their own experience with apostasy, the novelists inevitably impregnated their works with psychological realism. Maitland, Reade, Linton, Du Maurier, White, and Ward give every evidence of writing their own spiritual autobiographies. Consequently, these theological novels may lay claim to consideration as documentary evidence of the religious *Sturm und Drang* of the Victorian Age.

In the various novels considered evolution and natural selection, thrusting the heroes and heroines to the brink of disbelief and over, appear either as fiendish beasts or as angels of revelation. The theories variously destroy faith by presenting a satisfactory material explanation of nature's processes for an unsatisfactory supernatural one; by demonstrating the pitiless waste that marks the struggle for existence, and nature's lack of concern for individuals; by proving the six day creation a myth, the Bible a lie, and the Church a force of negation against science and progress; by the tide of evidence that demonstrates man is of the apes and not of the angels. Clergymen no less than devout churchgoers, driven by the theory, reach the same dead end of faith. The road to holy orders in the novel was a deadly one to those who dared not refuse the challenge of sceptical thought.

Consideration of the defection from faith will be divided between those novels which view the apostasy with sympathy—an overwhelming majority, and those which view it with disfavor.

I

In 1867 Edward Maitland published his *The Pilgrim and the Shrine,*[2] a tale of a hero's mental pilgrimage set in the framework of adventurous life in America. Maitland's novel of a pilgrim's progress aspired to do for English materialism what *Sartor Resartus* had done for German transcendentalism, representing the gradual change of a youth from a state of doubt to the final adoption of disbelief.

Inasmuch as the hero, Herbert Ainslie, a young Cantabrigian, began to have doubts when he took his degree in 1846, the novel has a decided post-Chambers, pre-Darwinian flavor. Indeed, all the evolutionary influences which ferment in his mind could have been derived solely from a reading of the *Vestiges of the Natural History of Creation* published two years before his graduation.

Ainslie, member of a religious and strictly orthodox family, is intended for the ministry, but his first scruples of conscience make him hesitate. Addicted to habits of abstruse reasoning, he finds inspiration in the "development theory" which sets him musing upon the "vestiges of creation." "Can monkey become man?" he queries, after seeing a simian imitate a negro gardener; but the realization of the moral and intellectual gulf between man and monkey leads him to discount the closeness of the relationship. There is no call for taking it by the hand with a "Hail, brother (or father) monkey."[8]

His dilemma having grown acute, Ainslie resolves to travel to settle his doubts, but the intended exile of a year becomes the exile of a lifetime. On shipboard he strikes up an acquaintance with an Episcopalian clergyman whose dualmindedness drives home to Ainslie the conflict between religion and science. In friendly discussion the Reverend Meade accepts unreservedly the doctrine of gradual development. He greets appreciatively Ainslie's attack on that apologetics which, in spite of evidences of countless ages of growth and change, insists on sudden creation at a specific moment. This Paleyan-Gosse conception Ainslie ridicules, suggesting that by such a doctrine it could be assumed that we all came into existence this very moment, equipped with readymade impressions in our minds which we mistake for memories of things past. Then, to Ainslie's consternation, the next morning the parson preaches from *Genesis* of the six days' work of creation with all the dogmatism of the unquestioning believer.

Inevitably, the young man's religious musings and scientific studies lead him to complete severance not only from the Church of England but from Christianity generally. He traces step by step the evolution of species and comes to feel that nowhere in nature, indeed, is there room for a God. The Catastrophists' belief that the world had developed by a sequence of convulsions might suggest the existence of God interposing His hand to redirect or recreate the world, but no one could give credence longer to the Catastrophic theory. The gaps in the geologic record were rapidly being closed up to prove that progress does not leap from one summit to another. So he rests in his evolutionary materialism, holding that given matter and motion he can account for all things, even for man.

Like Edward Maitland's *The Pilgrim and the Shrine,* Winward Reade's *The Outcast* (1875)[4] rings true, suggesting that both authors recorded their own spiritual odysseys on these fictional tablets. Extensive research in the literature of natural science destroyed Winwood Reade's own faith in the current Christianity of his time. To such study he was led by a devious route. Soon after the publication of Darwin's *Origin of Species* in 1859, the doctrine contained in it began to filter through from the learned to the general public. Among the many misrepresentations of it at that period was the statement unwarrantedly put in Darwin's mouth that man was descended from the anthropoid apes; and while the excitement produced by this misconception was at its height, Paul Du Chaillu, a Frenchman domiciled in America, exhibited in London three stuffed specimens of the gorilla, which he described as a newly discovered anthropoid ape of great ferocity and intelligence whose habitat was the forests bordering on the Gaboon, in West Africa. Hoaxes then as now were by no means unknown in scientific circles. The narrative of Du Chaillu, who was not fortified by any academic credentials, was at once assailed as a tissue of impossible lies, Dr. Gray of the British Museum Zoological Department leading off in May, 1861, with letters to newspapers headed "New Traveller's Tales."[5] Yet Du Chaillu found some defenders, including Professor Owen. Reade, whose tastes had early led him to the study of natural science, conceived the idea of visiting the Gaboon and deciding the controversy for himself. His journey to Africa in 1863 confirmed Dr. Gray's accusations of fraud. Again in 1868 Reade visited Africa to collect materials for a work *The Origin of Mind* suggested by *The Origin of Species.*[6] Darwin's *Descent of Man* anticipated this project; but Reade went on to publish his outline of universal history, *The Martyrdom of Man,* following in Darwin's footsteps "not from blind veneration of a great master, but because

I find that his conclusions are confirmed by the phenomena of savage life."'[7]

Reade died in 1875 at the age of 35 (his uncle, Charles Reade, wrote his obituary), but while the hand of death was actually upon him he penned his third novel, *The Outcast.*

The Outcast sets forth the persecution which awaited in the England of Reade's time the open profession of anti-Christian opinions. The book is in the awkward form of letters written by a dying father, Edward Mordaunt, to his daughter Ellen in which he describes with some pathos the struggles of an "intellectual" who, cast off by his family and driven from one employment to another on account of his religious views, sees his wife die of starvation, and is himself barely saved from the same fate. As a sort of prelude to this main action, Reade records the tragic fate of one, Arthur Elliot, friend of Mordaunt's, whose faith had been toppled over the verge by a ruthless science.

Religion was the poetry and passion of Arthur Elliot's life; the beneficence and benevolence of God, the rock upon which this religion was founded. But coming upon Malthus' *Essay on Population,* and Darwin's *Origin of Species* which it had inspired, Elliot devoured their contents. The diet proved too much for his devout soul. It unseated his reason. "In mourning for mankind," as he said, he dressed in black; and on his table placed *Malthus* lettered outside "The Book of Doubt," and *Darwin,* lettered "The Book of Despair." Shortly after, he committed suicide. Among his effects after he had hung himself was found a notebook containing an account of a number of nightmarish dreams induced by the reading of a *Quarterly Review* article on the Darwinian theory.

In his insane dream-vision "A New Thing Under the Moon," Elliott is transposed through infinity to the world of the demi-gods. With condescension the demi-gods occasionally turn their attention to the "poor mud-ball below," the Earth. One of their number has written a play dealing with this planet. The plot follows the Darwinian conception of the evolution of the earth and man. In the play the nebular hypothesis is brought to reality, then the evolution of the earth from a titanic "cinder enveloped in a cloud," then the origination of life kindled by the rays of the sun, the development of inorganic and organic species, and ultimately the emergence of man, from a jabbering tenant of the trees through various phases of civilization to modernity. In the Demi-god Club, Elliot comes upon a vitriolic review of this world-drama, in which the critic flays with a knotted whip the monstrous conceptions. The law of evolution—the plan of the piece—should have been modified, the critic insists. The persistence of the bestial in perfected man after the sloughing off of the animal form, the ruthless waste of life, death itself, meet with the demi-god's unmitigated scorn. "The law of evolution is the law of death . . . the earth is a vast slaughter-house," he sneers. The critic damns the God-author of this horrible creation, "a young beginner," and calls for a new work which "will give us a more favorable impression of the personal character of its Creator."[8]

Accidentally, Ellen, daughter of Edward Mordaunt, reads these revelations of a lunatic. The effect of the reading makes her, too, doubt the existence of a personal God. To offset this pernicious result with an antidote, Mr. Mordaunt tells the story of his life.

Spiritually, Mordaunt's life parallels in effect that of Arthur Elliott. As a young clergyman, unquestioning in his acceptance of the Bible as divinity, blissfully ignorant of the sceptical sciences, he fell a prey to the heresies of one, Fitzclarence, his rival in love.

"If it could be proved," Fitzclarence catechised Mordaunt, "that the world was not made in six days, but that thousands and thousands of years intervened between (for example), the fish and birds of the fifth day, and man who was created on the sixth, what may I ask would you say then?"

"Then," Mordaunt reasonably replied, falling into the trap, "of course it would be proved that the Bible was not inspired."[9]

The seed of doubt had been sown. Later by a ruse a copy of Lyell's *Principles of Geology* and some orthodox attempts at a reply were placed in Mordaunt's hands. He read them all, and his faith in the Bible was swept away. He resigned his living. Cast out of his father's house as an infidel, Mordaunt suffered a brain stroke, but was rescued by a Dr. Chalmers, scientist friend of the family. There, in Chalmer's house, a veritable College of the Sciences, housing in addition to the doctor, an astronomer, a geologist, and a comparative anatomist, surrounded by a telescope, rock-specimens, and the skeletons of the animal kingdom arranged systematically ("culminating with a chimpanzee and a man, standing side-by-side, their arms affectionately interlocked"), Mordaunt began a study of physical science and a long climb back to health.

Another clergyman's faith dissolves under the acid touch of evolution in James Franklin Fuller's novel *John Orlebar, Clerk* (1878).[10] John Orlebar's uncle, a bishop, deliberately buried alive the young man, a freshly made parson, in a secluded parish of Arderne, because of his dangerous views: John Orlebar was " 'unsound' on such questions as evolution."[11] Though he was ready to discuss evolution with the bishop, John Orlebar drew the line at preaching such doctrines from the pulpit. They would have been both embarrassing to the bishop and unpalatable to the parishioners. For all his Broad Church leanings, John Orlebar at this stage considered himself a sincere believer; and when the carping voice of Joe Twinch dared to cast doubt on his pretensions, the enraged hero put all the force of his muscular Christianity into a blow that almost cost Twinch his life. But the scientific view of nature fast gained ground on his old theological conceptions, and John in time concluded that it was unutterably base to hold a commission under false pretences. His decision to resign holy orders drew down on him the censure of his friend, Dr. Packenham. The rector, the doctor insisted, was deserting in time of danger. Consider the creed the world is coming to, he urged.

> Substitute Darwin for Moses, with a gorilla for his Adam. Let Huxley put on the mantle of Isaiah, and Tyndall, in the voice of Balaam, declare it impossible that an ass could speak.[12]

But John Orlebar would not be moved from his resolve; and intoned prophetically: "The old breastworks will not stand the on-coming assault."

Mrs. Eliza Lynn Linton, professed agnostic, friend of Herbert Spencer, and profound admirer of Darwin,[13] presents in *Under Which Lord?* (1879)[14] a novel with a saintly agnostic for its hero, and a High Church clergyman for its most objectionable character.

Richard Fullerton, the hero, has unseated Theology and enshrined in its stead, Science. His wife and daughter are indifferent to his philosophy and allow him to pursue untroubled his researches in a study where he is surrounded by skulls of chimpanzees and bushmen, astronomical plates, skeletons of fish, and photographs of the moon, an electrical machine, and a microscope. With the aid of these appliances he is accustomed to compose lectures designed to show the contradictions between nature and revelation, Genesis and science, to an initiated band of disciples consisting among others of the carpenter, tailor, and blacksmith of Crossholme, at an "Institute" built for the purpose.

But all is not plain sailing. The attitude of the Church, High and Low, towards science was in country parishes notoriously one of bitterest antagonism. The attitude of the Crossholme vicar, Launcelot Lascelles, is voiced through his sister:

> I hold all infidels to be possessed. They are the emissaries of the Evil One, and this so-called modern science is the means by which he works. But you will conquer in the end, Launcelot. The Church is stronger than the Pit.[15]

The vicar begins a systematic campaign to lay low the fortresses of unfaith in the parish. With jesuistical cunning he molds the plastic wills of Fullerton's wife and daughter to his own, and secures the aid of one of Fullerton's pharisaical disciples "loudest in his admiration of the doctrines which developed him from a jelly-fish through the common progenitor of himself and an ape."[16] Richard continues to lecture to the workingmen on the impossibility of reconciling Genesis and Geology, but underneath his feet his Institution is crumbling. Under the clergyman's tutelage the Fullertons begin to look on Richard as one vomited from hell. "You believe in nothing at all except your horrid old bones and senseless bits of jelly," Mrs. Fullerton sneers;[17] and daughter Virginia, coming from a Christmas feast informed by high Christian art and aesthetics, to his laboratory of heathen science, turns from her father in revulsion. "He is an atheist," she says, in a voice of horror, crossing herself as she speaks.[18]

The astute Lascelles harries Fullerton further by inviting orthodox popularizers of science to lecture at Crossholme on Fullerton's special Mondays at the Institute. But his greatest triumph in the fight against the Crossholme Apollyon occurs when he succeeds through legal machinations and Mrs. Fullerton in taking the Institute out of the agnostic's hands.

Filled with tragic bitterness, Richard Fullerton delivers his last lecture, his swan-song of agnosticism, tracing "the development of man from an ascidian and the close chain of likeness running through the whole race of the vertebrates."[19] But like the *"Delenda est Carthago"* of the Elder Cato is the climax of his discourse, his old argument against spiritual infallibility, concluding,

> So soon too as the doctrine of evolution in nature became established as a scientific fact, true in substance if in parts faulty in detail—so soon did the Bible become a simply human record of puerile fables mixed up with lofty thought—interesting as an historical study, but a dead letter as Revelation.[20]

Shortly afterward Fullerton dies. His daughter Virginia becomes a nun. Having bled Mrs. Fullerton of all he can, Lascelles turns away from her. She suffers a violent reaction. Months later she returns to the Institute stripped bare now of sinful evidences. To her it is a mausoleum.

> How she would have welcomed even a hideous skull, or the godless, soul-destroying portraits of a nosed ape and a flat-faced savage set side by side as shameful parallels.[21]

In *The Autobiography of Christopher Kirkland* (1885),[22] Mrs. Linton used the same sympathetic colors in painting the portrait of the sceptic Christopher as she had in limning Richard Fullerton. But the career of Kirkland parallels more nearly that of Herbert Ainslie, hero of *The Pilgrim and the Shrine,* and like that work charts its course by the milestones of evolutionary history.

Christopher Kirkland passes through a Centre of Indifference to the Everlasting No after ringing the changes upon Unitarianism, deism, spiritualism and several other varieties of religious experience. In his youth his father intended him for the ministry, but that hope was shattered on the *Vestiges of Creation.** To Christopher that

* The *Vestiges* likewise undermined the faith of the Reverend Ernest Pontifex in Samuel Butler's *The Way of All Flesh* (1903). This was before the days of Evolution. After a talk with the freethinker, Mr. Shaw, and an afternoon at the British museum with Chambers's work, Ernest was ready to jettison both his Bible and his morality.
Samuel Butler, *The Way of All Flesh* (New York, 1929), pp. 282-284.

book was a Bible, to his father it was a Demonology. Convinced of the truth of Christianity and the perfection of Anglican Protestantism, the elder Kirkland tries to save Christopher's soul, but in vain. They part.

The worldly-minded Christopher then sifts through several philosophies, religions, and speculations like Pater's Marius, but finally strikes bottom in the firm soil of science. He mixes with the scientists and follows eagerly the development of the evolutionary theory as Chambers, Darwin "the epoch-maker and torch-bearer of this century," Lyell, Huxley reveal its facets; he is a prejudiced spectator as Owen and Huxley battle over the *Hippocampus minor.*[23] Darwinism has opened a new world before him.

Evolution becomes his solace and his creed. In a scientific evolution of morals, systems of government, education, and laws of health he sees hope for society; and in the evolution of man's mind by the same laws which "produced brain from protoplasm" he sees hope for man himself. Evolution then to him is guide, philosopher, and friend, a Law of Progress that suggests that "just as from the lowest forms of life has been evolved man, so from the brutality of primitive communities where the stronger kill the weaker" have been evolved law, cooperation, and conscience, the social instincts.[24] "When Darwin died he had lived. He had fulfilled his appointed mission, and planted his Tree of Life fathoms deep in the soil of human thought and knowledge."[25]

Mrs. Linton drew heavily on her own spiritual experience and remembrance of things past in equipping Christopher Kirkland's mental larder. Born in 1822, she came to maturity when the earth was filled with signs and portents of the coming evolutionary storm, and when Darwinism broke upon the world in 1859, she was thirty-seven. George Gissing, on the other hand, born in 1857, grew to maturity in a generation which reluctantly and somewhat hazily compromised with the Darwinian theory, seeking only to prove "that 'The Origin of Species' was approvingly foreseen in the first chapter of Genesis."[26] Consequently, Helen Norman of Gissing's *Workers in the Dawn,* an intelligent sceptic, begins the undermining of her faith where Christopher Kirkland left off.

George Gissing's father was an enthusiastic botanist and compiled a *Wakefield Flora,* but Gissing himself detested the sciences. Nevertheless, by the time *Workers in the Dawn* (1880),[27] his first published novel, was being written, Gissing had made himself familiar with the development of nineteenth century thought, and had reached definite conclusions, religious, social, and philosophical. The chapter "Mind Growth" in *Workers in the Dawn* contains, in Helen Norman's diary, what may probably be accepted as a substantially accurate record of Gissing's own studies in his effort to find and place himself intellectually.[28]

Inspired to honest doubt by reading Strauss, Helen Norman travels to Germany to continue her studies in scepticism. Though she heard her father speak of it, she has never read Darwin's *Origin of Species* when one day in Tübingen she comes upon her instructress reading a German translation of the work. Helen devours its contents "with an enthusiasm . . . such as no other book except the 'Leben Jesu,'" ever stirred in her. In the *Origin* she finds strong support for her disbelief, and she prepares to make a religion of natural science. We learn that Darwinism

> . . . is a theory which recompenses me a thousandfold for my loss of the old Biblical superstitions. . . . Yes, yes, this is real, solid food, no insubstantial cloudshape or chimera. Here is a theory built up on solid facts, facts one can grasp, handle, sample with the eye or the microscope.[29]

Miss Norman pursues her studies, reading greedily Haeckel's *Natürliche Schöpfungsgeschichte,* a work inspired by Darwin, and Comte's *Philosophie Positive.* Entranced by positive law as reflected in these works, she exclaims, "Comte is for me the supplement to Darwin; the theories of both point to the same result, and *must* be true."[30]

The deadly seriousness of *Workers in the Dawn* is in distinct contrast to the mock-seriousness of Richard Dowling's *Under St. Paul's* (1880),[31] of the same year.

The hero, George Osborn, a wide-eyed rustic from Stratford-on-Avon, falls in love with Marie Gordon, as forward as he is shy. But her flippancy changes, and under the devout Osborn's guidance she even becomes an orthodox member of the Church of England; but he is not satisfied until she takes a solemn vow she will marry no man who does not belong to the faith of the Church. This last provision, Osborn hopes, will prevent Marie from backsliding in the event of his death. Unhappily, however, he himself loses his faith and thus becomes ineligible to marry his beloved.

His conversion to disbelief is gradual. Confident in his faith, he accepts an invitation to a meeting of the Prehistoric Society where he meets "the illustrious broacher of the chimpanzee theory,"[32] and hears a lecture that is anything but orthodox. His friend Nevill lends him some books on natural history. He reads them. "What monstrous blasphemy!" he exclaims. "Man the accidental descendant of the ape! Why is not this book burned by the common hangman?"[33]

A few days afterwards he finds himself at the Zoological Gardens. One of the keepers offers to show him some curious animals. He mistakes them for negroes.

> "In the name of God, then what are they?"
> "Chimpanzees."[34]

About him lie the ruins of his old faith. About those ruins, in his mind's eye, he sees leaping and bounding ten thousand forms of loathsome brutes, grinning at him, grimacing. "They took up the relics of that sacred palace, that holy fane, and smashed and tore and cast them about." And in his ears resounds the din, "The reign of the Beast! The reign of the Beast!"[35]

He rushes out of the Gardens, trailing the remnants of his faith. Of course he cannot keep the heroine to her promise to marry him; but everything does turn out happily at the end.

Like the heretic hero of Edward Maitland's *The Pilgrim and the Shrine,* Mark Rutherford, hero of William Hale White's *The Autobiography of Mark Rutherford* (1881),[36] is educated with a view to becoming a minister. He attends a Dissenters' college where he receives an overdose of Calvinistic theology. But the reaction does not set in until after he has been ordained. Realization of the implications of the Darwinian hypothesis plays a part in destroying his faith.

Calvinism had taught him God's concern with individuals. But Darwinism taught him that this was not nature's way. He cannot foresee an immortality for man who though produced millions of years ago is still a creature "scarcely nobler than the animals he tore with his fingers." For the law of the universe everywhere "is rather the perpetual rise from the lower to the higher."[37] Imperfect man surely will be thrown into discard in the ruthless struggle for existence. But Rutherford's negation is not complete; with Browning he holds grimly to the belief that this ceaseless struggle must have a meaning.

Like *Mark Rutherford,* Mrs. Humphry Ward's novel *Robert Elsmere* (1888)[38] contains authentic autobiography; like William Hale White, Mrs. Ward set herself to convey to the vast public who read little else but fiction the message that dogma is not essential to true religion.

Robert Elsmere is a product of the intense spiritual unrest whose atmosphere Mrs. Ward as Mary Augusta Arnold breathed at Oxford from 1865 to 1881. Round about her were whistling the winds of an agitated day. She came there a girl of fourteen when the University was the center of the rising Broad Church school, and was permeated with the spirit of Darwin's *Origin of Species* published six years before—both impugning the tradition of the Semitic Scriptures.[39] A bride at the age of 21, Mary Augusta Arnold lived at Oxford for nine years as a young married woman, following the special bent of her mind in application to what she herself calls "the general literature of modern religion." Related to the Huxley's through the marriage of her sister Julia (Mrs. Leonard Huxley) who was her most intimate friend, Mrs. Ward followed with pardonable zeal the course of modern scientific inquiry, especially as it affected religion.

Like Edward Mordaunt of *The Outcast,* Robert Elsmere, hero of Mrs. Ward's novel, is a clergyman, devout, credulous, destined to pass through an ordeal by fire for his religious views. Insulated by his romantic religious passion, Robert passes through Oxford to Holy Orders with his beliefs intact. But in Murewell parish where he has settled, he finds leisure to conduct those studies and researches which have always been near his heart, and soon, unwitting, he is on the high road to unfaith. Scepticism, impersonated by Squire Wendover, Murewell's landlord, introduces Elsmere to the Higher Criticism, and the clergyman's fate is sealed. Finally, only one shred of faith is left him: belief in the Living Christ.

Relations with his militantly sectarian wife are strained to the breaking point. Finally, as had Mordaunt, he resigns his living, cuts himself aloof from the church. His scarified spirit seeks solace everywhere, and finally comes to rest in an East End settlement. Where Mordaunt in Reade's novel found ultimate balm in Positivism, Mrs. Ward's Elsmere discovers a spiritual anodyne in an ethical culture that retains the Christ but doffs the creed; and at his death he leaves a monument to his religion, "The New

Christian Brotherhood," a Christianity emptied of the lumber of theology.

The truly devastating blow that splits Elsmere's faith in twain is delivered by his own historical researches, and by the telling arguments of the apostate Squire. But it is Darwinism which drives in the opening wedge.

The clergyman actually flushes with embarrassment when the critical Langham, his former tutor, discovers him at Murewell reading Darwin and Grant Allen; but not for long does he flush.

"Imagine, Langham!" he said presently, "I had never read even 'The Origin of Species' before I came here. We used to take the thing half for granted, I remember, at Oxford. . . . But to drive the mind through all the details of the evidence, to force one's self to understand the whole hypothesis and the grounds for it, is a very different matter. It is a revelation."

But the voice of doubt, mocking his enthusiasm, begins its work. "It is a revelation," Langham echoes; then adds significantly, "that has not always been held to square with other revelations."[40] Robert, for all his "driving" and "forcing" of mind, seemed to have shut his logical sense to the implications of Darwinism for theology. He recognized that the new facts were "troubling," had caused indeed a panic among theologians, but "We are not saved by Darwinism . . .," he announces airily. "The old truth remains the same. Where and when and how you will, but somewhen and somehow, God created the heavens and the earth!"[41]

But he dare not press the facts of evolution on his wife. It would distress her. So Elsmere is made aware for the first time of the difference between his and his wife's beliefs; for the first time he is turned to consider his orthodox faith in a critical way. Soon that which suited a part infects the whole. His scientific work, fragmentary as it was, begins to quicken and sharpen the comparative instinct. "Evolution—once a mere germ in the mind—was beginning to press, to encroach, to intermeddle with the mind's other furniture."[42] The way had been made clear for the demon of unbelief.*

Robert Elsmere was an enormous popular success.[43] In two weeks four editions were exhausted; and within the year hundreds of thousands of copies were circulating in sixpenny and sevenpenny editions. The theological novel, hurling defiance in the face of those who gloried in "art for art's sake," had come into its own. The phenomenal interest of the public in the soul-struggle of a clergyman was more than evidence of the popularity of a book; it was proof positive of a curiosity approaching the insatiable concerning the religious issues of the age. Robert Elsmere's dilemma was The Lady or the Tiger? of the 'nineties.

A veritable dragon's tooth Mrs. Ward's famous novel proved itself. After *Robert Elsmere* the novel became entrenched as the thing wherein to catch the conscience of anyone not possessing the proper amount of orthodoxy or heterodoxy. A host of novels patterned after Mrs. Ward's story of religious doubt, sprang into being. The Hero as Clergyman, Elsmere in different guises, played the stellar role in many of these. Representative is Joseph Hocking's *Jabez Easterbrook* (1890).[44]

Jabez Easterbrook, a lesser Elsmere, is an ideally good, intelligent Methodist preacher. Easterbrook falls in love with the daughter of his elder or principal church member, a haughty, beautiful young atheist named Margaret. On first hearing of Easterbrook, Margaret makes a sneering remark about sanctified Methodists, then turns to Darwin's *Descent of Man* "for a little sensible society."[45] She rather

---

*Delia Blanchflower, heroine of Mrs. Ward's novel of that name (1915), succumbed more easily than Robert Elsmere: "A few stray books or magazine articles had made a Darwinian and an agnostic of her."
Mrs. Humphry Ward, *Delia Blanchflower* (New York, 1915), p. 61.

opens Jabez's eyes by telling him his religion is all nonsense and that if he studied geology and evolution he would find it out. But the hapless Jabez who still believes in a literal interpretation of the Bible with regard to creation, and that the human race is merely 6,000 years old, cannot believe her words. Dogmatically she assures him no God is necessary for creation: that in the chemical laboratories the gap between the inorganic and organic world has been bridged.[46]

Easterbrook spends a year—or that part of it which he could spare from knocking down drunken navvies and converting them into teetotalers—in reading books on science. But he is not convinced; if anything he becomes a better Christian than ever. "I have come to see this," he remarks, "that all scientific discoveries are revelations from God";[47] and proceeds to incorporate his knowledge of those discoveries in sermons. But his studies have destroyed his belief in fixed creeds; and soon he finds himself charged with heterodoxy. Most damning is the accusation that he has fraternized with a Unitarian minister, together with him engaged in free discussions of the truths of the Bible and read "not soul-saving books, like the Life of William Bramwell, but books by Darwin, Huxley, Spencer, and Harrison."[48] Dismissed from the ministry, he marries Margaret who by this time has become as good a Christian as he, and sets up in a neighboring town as a Dissenters' Robert Elsmere.

Loss of faith was a serious matter to the Victorians. In *The Outcast,* in *Robert Elsmere,* and in *Under Which Lord?* it was presented with a *Weltschmerz* leitmotif and the trappings of tragedy. But occasionally as in Dowling's *Under St. Paul's* and in Du Maurier's *Trilby* (1894),[49] the theme was decked out in motley. This, in the case of George du Maurier, was characteristic. Whether wielding the brush or the pen, Du Maurier was always puckish; and even in limning the clash of evolution and religion as it might reasonably have affected his own life, he mixes the bitterness of a Jaques with the sweet humor of the Clown.

Du Maurier presents the problem of Little Billee Bagot "whose only choice lay between Mr. Darwin and the Pope of Rome,"[50] and finds himself in the unenviable position of suitor for the hand of a devout daughter of a still more devout clergyman—all this in the early 'sixties, "long ere Religion had made up her mind to meet Science half-way, and hob-knob and kiss and be friends."[51]

Little Billee Bagot lost his 'ane true love' when Trilby disappeared. He did not know that Mrs. Bagot, alarmed at her son's betrothal to an artist's model, had induced Trilby to run away lest the girl ruin his budding career. Broken in spirit and weakened by a disastrous illness, Billee is easily prevailed on by his mother to cast his eye elsewhere, on sweet Alice, his sister's friend. A dream, working on his fevered mind, decides him: he will ask Alice's father for her hand in marriage.

Billee is in a precarious state. The *Origin of Species* (which he is reading for the third time) has proved "a little too strong for him"—it has unseated his faith in the "good old cosmic taradiddles" of orthodox theology. But, like Robert Elsmere, not for the world will he seek to undermine the faith of a pure woman; for sweet Alice still believes everything she has been taught; even in the traditional cosmology of a

> wrathful, glorified, and self-glorifying ogre in human shape, with human passions, and most inhuman hates—who suddenly made us out of nothing, one fine day—just for a freak—and turned us adrift the day after—damned us from the very beginning—*ab ovo—ab ovo usque ad malum* . . . and ever since! never gave us a chance![52]

Little Billee cannot bring himself to "pop the question" to Alice's parson-father. He is deadly certain of being quizzed on his articles of faith—which he has, of course, dumped overboard; and with them, he expects, his eligibility as a son-in-law. In his dilemma he apostrophizes

Alice's dog Tray who offers tail-wagging encouragement, but no advice. Finally, Billee decides to cheat—to lie to the parson about his state of grace. In the midst of his expostulations to Tray, the vicar comes upon him.

"What book's that you've got in your hand, Billie?"

"A-a-a it's the *Origin of Species,* by Charles Darwin. I'm very f-f-fond of it. I'm reading it for the third time. . . . It's very g-g-good. It *accounts* for things, you know."

And Little Billee proceeds to blurt out a confession of his unfaith until, white in the face, the parson splutters with rage, "Sir, you're—You're a—You're a *thief,* sir, a *thief!* You're trying to *rob me of my Saviour!*" And then comes the thunderous Victorian refrain: "Never you dare to darken *my* doorstep again!"[53]

The infallibility of the Catholic Church, and its avowed stand against evolution were no more sufficient to stem the destructive tide of evolutionary doctrine than the Canute-like gesture of the Anglican Church. Indeed, in its fanatical insistence on the very letter of the miraculous revelation, Catholicism invited heresy.

The Roman Church frowned puritanically on the opera and the stage, but Evelyn Innes, in George Moore's novel of identical name,[54] nourishes dreams of exhibiting her talent for singing outside the church choir of Dulwich town. Good Catholic though she is, she cannot resist the temptation to accept the offer of Sir Owen Ascher, a sophisticated Londoner, to accompany him to Paris to study music. The accomplished seducer makes her his mistress, but fearing that her religion may revive in her moral scruples, he drenches her mind with evolutionism. In the course of a meteoric career she falls out of love with Sir Owen and determines to quit the stage and return to the religion from which Owen seduced her. Protracted is her struggle to "work out the beast" and receive absolution for her sins; but the patience of Monsignor, a Roman prelate, helps her to a weak success.

It is during Monsignor's catechizing of Evelyn to prepare her for absolution that we learn the details of her retreat from faith. Under the tutelage of Sir Owen, himself an avowed atheist, Evelyn was put through a thorough course in Huxley and Darwin, with a grounding in Herbert Spencer.

That Spencer should have been included in the curriculum provokes the Monsignor. Significantly ignoring the other two, he vociferates: "Herbert Spencer! Miserable collections of trivial facts, bearing upon nothing . . . the general law! rubbish, ridiculous rubbish!"[55]

Evelyn read evolutionary science not in the least to obtain knowledge, but to undermine her belief in God. But that belief has sunk deep into her consciousness, too deep to be easily rooted out. In spite of her desire to convince herself, remembrance of Owen's arguments and all that she had read in Huxley and Darwin whistle like a wind through her pious exhortations; "We know nothing . . . we know nothing . . . ," she heard in the shriek of the wind, and revealed religion appeared in tattered miserable plight, a forlorn spectre.[56]

Monsignor is clever. If her scientific knowledge is the stumbling block in the way of her faith, he would remove that. Sophistically he proceeds to prove that the teachings of Darwin, Huxley, and Spencer which have inspired her to this sterile liaison with Sir Owen thwart the designs of nature. "But," Monsignor concludes, "if we go to the root of things we find that the law of the Church coincides very closely with the law of Nature, and that the so-called natural sciences are but a nineteenth century figment."[57]

Evelyn acquiesces. Finally in her ears sounds the *Ego te absolve,* and "in that supreme moment for which she had longed, the last traces of Owen's agnostic teaching seemed to fall from her."[58]

II

On the other side of the ledger are those novels which reflect the unmitigated disdain of the author for the theory which dared call into question the authority of Revelation, or the beneficence and even the existence of God. As in the case of novels attacking evolution or Darwinism directly, there is a notable decline in the number and rancor of works of this order after 1880. After that only Marie Corelli continued to drip venom.

The novels here represented portray the atheist as a profligate beyond redemption, a wretch who "with studied, sly, ensnaring art" betrays unsuspecting youth into disbelief, or diabolically spreads the seeds of infidelity. They depict the atheist's retreat from apostasy, convinced by miraculous revelation of the truth of God; or stress the tragedy of life without God in comparison with the bliss of life with God.

Julia Kavanagh, the obviously orthodox author of *John Dorrien* (1875),[59] has only disdain for the freethinker. In the novel, Oliver Black is limned as the conventional beast-atheist. A smooth-tongued handsome scamp, Oliver seduces into unbelief the young Antoinette who "had a good fund of ignorance for him to work upon." Antoinette listens with rapt attention to what the author calls Oliver's "pantheistic" doctrines, his ungodly description of a universe which no God had ever created, "a world ever developing itself in vast unbroken progress."[60] Finally, Antoinette is saved from atheism by casting off her heathen lover and becoming the lover of a pious young Christian.

Once again, as in *John Dorrien,* in *That Artful Vicar* (1879)[61] by E. C. Grenville-Murray, a profligate pyrrhonist is set as a foil against a man of God, the one "a scoffer flushed with wine . . . the other a believer, sober, grave, and bearing on his features the impress of the Ten Commandments."[62] The inevitable clash between the two occurs at the dinner table, a staple article in all Victorian religious debates.

The wealthy *giaour* Sir Giles Taplow provokes hostilities by baiting the exemplary young vicar of Stillborough, Paul Rushbrand, with a flippant disquisition on "pre-Adamite" discoveries of old flints, "considered a 'facer' for theologists."[63] With malice in his voice the squire announces his intention of sending Darwin's *Descent of Man* "which traces us all up to baboons and frogs"[64] to the Mechanics Institute at Stillborough that the poor may learn. The curate leaps to the challenge. Did Sir Giles dare to deny the existence of God or divine creation on the testimony of such a book. Did Sir Giles dare to ignore the testimony of the beauties and harmonies of nature as evidences of a designing hand? Yes, Sir Giles dared. But Paul is not nonplussed. He is not ready to admit that Darwin's book is scientifically sound, even if it were demonstrated

> . . . that atoms had combined to form, first the sun, then the earth—that air working upon earth had produced moisture, that from the moisture the sun had drawn vegetation, and vegetation had bred animal life—first the mite, then the newt, the tadpole, the frog and so on through the ape up to man.[65]

Even if all this were proved, Paul insists, there would still remain the question, who created the first atoms and the air which vivified them? If Sir Giles was ready to admit that the primary particles had to be created, then he would have to admit that the power which could create those particles which evolved, was quite as competent to build at once the entire world. Since there was no effect without a cause, Sir Giles must, surely, admit the First Cause?

But the game of "proving the Divinity by a series of logical sequences" comes to an abrupt end when Sir Giles slips from his sleeve a trump card—soon to be worn threadbare—by querying, "Then who created the Creator?" Paul retreats in a cloud of words about mysteries which God intends to remain so. Then humbly he entreats Sir Giles not to sow the seeds of infidelity among "our lowlier brethren" by giving them access to such a book as Darwin's *Descent.*[66]

Considering the sympathies of the author, it is strange indeed that the devout parish priest should have been worsted in this passage of arms. But certainly by over-zealousness in the cause of faith Paul Rushbrand invited his defeat. By accepting the squire's premises, and arguing along irreligious evolutionary lines himself, Paul made a fatal mistake of which the sophistical Sir Giles was quick to take advantage. Worthy of remark, however, is the exposition of the materialistic concept of evolution which Paul is made to deliver—in its comparative accuracy rare among those to whom the term "atheist" signified everything loathsome.

Shortly after the publication of *Robert Elsmere,* Marie Corelli, already a famed writer through the success of her fantastic *A Romance of Two Worlds* (1886), began a novel intended to answer Mrs. Ward's *chef d'oeuvre.* This was *Ardath*[67] which appeared in 1889, a year after *Elsmere.*

Marie Corelli considered it her special providential mission to preach against unbelievers, to put to flight the armies of Tyndallism with a barrage of words. "Atoms" and "molecules" specifically call forth her scorn, but while damning the atomic theory (as conceived anew by the author), Miss Corelli put forward an electric theory of her own, a doctrine of soul-germs and electricity protoplasm that defies religious classification. "God is a shape of pure Electric Radiance," she tells us in *A Romance of Two Worlds;*[68] and again, "Cultivate the Electric Spirit within you."[69] To this conception of the essential Christianity, Marie Corelli added an "Electric Origin of the Universe." This theory, she insisted, scientific professors, who were forcing on public attention "ingenious explanations of 'atoms' and 'molecules,'" would in time acknowledge as "the only theory of Creation worthy of acceptance."[70]

In her novels Miss Corelli attempted to combine the substance of imperfectly comprehended contemporary science or pseudo-science with a mystical philosophy, a Christianity distorted beyond belief. This strange amalgam the novelist treated as revealed from heaven; whatever in Christianity or science ran counter to it she pursued with a flood of words and a whirlwind of high-sounding and empty syllables.

Such a cloudburst drops upon the atom first, then upon *Robert Elsmere,* and finally upon Darwin in the romance *Ardath.* Therein Marie Corelli represents a hero who, infected like Elsmere with an apostate virus, is converted back to spiritual health. Theos Alwyn, the hero, grounds his atheism and disbelief of the soul on the fact that science can give no positive proofs of the soul's existence.* His science, however, is of a nebulous and unstable order. With remarkable ease he is set back on his heels by the salvo of questions the monk Heliobas fires at his proclamation of disbelief. Answering Heliobas's query as to how science can explain existence, Alwyn responds vaguely,

"Evolution and Necessity."

"Evolution from what?" persisted Heliobas. "From one atom? What atom? And from whence came the atom? And why the necessity for any atom?"

"The human brain reels at such questions," said Alwyn, vexedly and with impatience. "I cannot answer them—no one can!"[71]

---

* Earlier, in *Thelma* (1887), Miss Corelli represented such a character in the person of Lorimer, whose religion was knocked out of him by a 'scientific professor.' Lorimer believes in nothing: "I came from nothing —I am nothing—I shall be nothing."

Marie Corelli, *Thelma* (New York, 1906), p. 66.

Under the inspiration of Heliobas, Alwyn then passes through a mystical experience—the vision of an angel—which effects in him a regeneration of faith. In an ecstasy of religion he composes a poem "Nourhalma" which makes him a literary celebrity. His friends rejoice at his conversion to the faith, but his old mentor, Professor Moxall, an evolutionist, mourns the loss of his most promising disciple in the material sciences. "Why, with such a quick analytical brain as his, he might have carried on Darwin's researches to an extremer point of the origination of species than has yet been reached."[72]

Alwyn's new faith, "no more original than the doctrine of evolution," suggests vaguely a fusion of "the 'Divine Spirit' of the Christian religion with the advancing inquiry and scientific research of man."[73] At a reception given by a duchess, he expounds his credo to the amazement of all, and then joins with the party in a general tirade against the novel *Robert Elsmere*. Pulling out all the stops, Alwyn lets out a blast of invective against the uprooter of faith guilty of "as gross an act, as that of one who loaded with benefits scruples not to murder his benefactor."[74] So effective is his evangelical ardor that the Duchess, atheist wife of an atheist Frenchman, begins to babble of religion and a "better life," and denounces the lack of faith in the world.

Once more Miss Corelli came down in sledge hammer style on the unoffending atom, and additionally on the fortuitous element in evolution in the novel *The Mighty Atom* (1896),[76] which she sarcastically dedicated to the "Progressivists" who believe in

> . . . that wonderful First Atom, which without knowing in the least what it was all about, and nobody to guide it . . . produced such beautiful creations.[77]

It was by denying to precocious eleven-year old Lionel Valliscourt the consoling belief in God, and by cramming his head with Corellian misconceptions of the nature of the atom and its part in creation, that Lionel's atheistic father drove the boy to strangle himself to death.

A Mr. Skeet showed little Lionel the enlarged drawing of an "atom" as seen through a microscope, "a curious, twisty thing with a sort of spinal cord running through it" and explained to the impressionable child that it was "a fortuitous combination of such things that made universes."[78]

Obviously an atom which was marked down its back by a spinal cord would be no atom at all, but an organism, hence subject to division. But here as elsewhere Miss Corelli's science is of the vaguest; but it was necessary for the author to set up straw theories that she might the more easily knock them down. Both Mr. Valliscourt and Professor Cadman Gore, Lionel's teacher, who are the defenders of the atomic theory of creation, in common with all of Miss Corelli's "brood of atheists," are confounded with utmost ease. The Professor, in fact, is overwhelmed with confusion when submitted to Socratic questioning by the youthful Lionel, and ultimately is restored to faith.

Early in the struggle between Scripture and Science, there were some on the side lines who called for conciliation of differences, and others who urged on the fight with no quarter given. There were those, of course, like the clergyman Meade whom Herbert Ainslie of *The Pilgrim and the Shrine* (1867) met on board ship, who achieved a seeming concord by reserving for private conversation one set of views, and for the pulpit an altogether different set; but this duality, however feasible, struck many as insincere. Other clergymen, in their devoutness, shut their eyes to all the implications and like Jabez Easterbrook intoned a benediction over the new science: "All scientific discoveries are revelations from God." But as evolution swept over barricade after barricade which the

Church had erected against it, theologians sought more anxiously to make terms with the conqueror before he should lay waste the last citadel. Richard Fullerton, the atheist of *Under Which Lord?* (1879), made a disparaging reference to this growing activity of the Church which sought "to make it plain that Genesis demonstrates the truth of geology and that no Christian scientist need trouble himself about the physical cosmogony of the Bible."[79] And even Marie Corelli was ready in one of her cloudy pronunciamentos to urge religion to "accept, absorb, and use the discoveries of Science."[80]

These random suggestions of compromise between religion and science became more and more frequent as the century turned its third quarter.

CHAPTER EIGHT

Compromise and Conciliation

The influence of evolution upon theology presents itself in a threefold aspect: first as destructive, second as transforming, and third as constructive. When evolution was first offered as an explanation of the world its destructive aspect as regards theology was about all that a considerable element in the Church could see. Here was an interpretation of things that was nothing less than a flat contradiction of revealed truth. It seemed to strike at the roots of a belief in God as the Creator of the world. It assailed the cornerstone of theology—the fall and total depravity of man, and in its materialistic form, seemed to extinguish all religion. Men forged, we have seen, no end of hastily constructed and easily demolished arguments against it; or refused to discuss the matter altogether, seeming to go on the principle that if they ignored the situation it might go away. But the more patient among them, realizing the solidarity of the scientists' position, studied and tried to understand evolution's bearing upon what the religious world had hitherto held as truth, and it was seen to have many helpful outlooks. Soon, in a remarkable *volte face,* and with hardly a reference to its former hostility, the Church consented to make terms and to live in at least outward harmony with its former foe.*

The war of ideas passed from the first to the second phase.

From all sides came evidences of desire to square dogma with the newly discovered laws of nature. Strict adherents of the Biblical texts pointed significantly to the verses in

---

*To the statement that a Roman cardinal was giving lectures on science, Apollonia, in The Earl of Beaconsfield's *Lothair,* replies:

It is remorse. Their clever men can never forget that unfortunate affair of Galileo, and think they can divert the indignation of the nineteenth century by mock zeal about red sandstone or the origin of species.

Earl of Beaconsfield, "Lothair" (in *Novels and Tales,* vol. X), London, 1900, p. 31.

Genesis in which the earth and sea are made to bring forth birds and fishes, and man was created out of the dust of the ground. Others, no less orthodox, began to disclaim the Church's intention of treating the Bible as a treatise in geology, zoology, or astronomy; or of taking the story of the creation in six days in a too literal sense. The apologetic disposition of the 'eighties was so strong that, faced by the alternative of believing either the Genesiac account of human creation from the dust or the Darwinian account of descent from the brute ("modified mud or modified monkey" as one of Darwin's reviewers strongly put it), the clerical world in large numbers was ready to follow Darwin's lead. Man, formerly "the reformed ape," was now viewed as an "undeveloped angel" striving upward from mean beginnings towards a goal unrealized. Broad churchmen, generally, took ground in Darwin's favor, realizing in Darwinism for the first time a new argument for design in Nature. Men were told that "through the ages one increasing purpose runs," and that evolution was the unfolding of an idea which had been in the mind of God throughout. In this they were merely treading the path which the Reverend Charles Kingsley, natural scientist and muscular Christian in one, had walked many years before.

I have gradually learned to see (Kingsley wrote Darwin in 1859 on receipt of a presentation copy of the *Origin of Species*), that it is just as noble a conception of Deity, to believe that He created primal forms capable of self-development into all forms needful *pro tempore* and *pro loco*, as to believe that He required a fresh act of intervention to supply the *lacunae* which He himself had made. I question whether the former be not the loftier thought.[1]

Once it had been shown that what Huxley called the "Miltonic hypothesis" of special creation was untenable, and Darwinism ceased to draw fire from churchmen of every faith, Victorians welcomed the new theory as a bringer of glad tidings. England was growing richer year by year, and enjoying budget surpluses in spite of diminished taxation. Her machinery was the wonder of the world and her parliamentary government a model for the imitation of enlightened foreigners. No one could doubt that progress would go on indefinitely. In such a world, impregnated with the sense of material progress, evolution seemed only a generalization of everyday life; and the concept of gradual change, gradual progress, especially suited the British temper.

Even the suffering implicit in the struggle for existence was extenuated in this theistic attempt to justify the ways of God to man. The struggle for life was interpreted as an instrument for perfecting the race. A larger good is evolved through the winnowing process by which physical nature casts its weaker products aside. Complacently, the successful business man adopted the doctrine of the "survival of the fittest," and now found justification for rugged individualism and *laissez-faire* in natural law.

Darwinism had won its way to respectability. Divines sidled up toward the object which at first had struck them as something spewed from hell, and discovered that after all it was not so diabolic a phenomenon as they had imagined. Its breath did not wither up every lofty aspiration and every worthy conception of the destiny of humanity. Evolution, once repelled, rejected, persecuted, shrieked at as the destroyer of souls, had lost its hoofs and horns, and the world—even the clerical world—settled down in the belief that it simply reflected the truth of Nature; that we who are now "foremost in the files of time" have come to the front through almost endless stages of promotion from lower to higher forms of life.

Thomas Huxley lived to attend another meeting of the British Association at Oxford in 1894, at which Lord Salisbury, the leader of the Conservative party and Lord Chancellor of the University, declared as President of the Association that the theory of evolution had been accepted, universally, by the scientific world. Huxley wrote to Hooker: "It was queer to sit there and hear the doctrines you and I were damned for, thirty-four years ago, enunciated as matters of course, disputed by no reasonable man."[2]

The movement to show that the lion might lie down with the lamb in perfect peace and understanding asserted itself with increasing fervor in the 'eighties and 'nineties. Just as in the first and second quarters of the century the Coleridgeans were able to reconcile Faith and Reason by the higher way of the Understanding, so in the fourth quarter certain members of the Broad Church especially, strove to reconcile Genesis and Darwin by the elastic theory of Development. The step was far from a unanimous one. Zealots in both the religious and scientific camps held back, unwilling to be swept along by the onrushing tide of compromise. And others, who by no interpretation can be labeled die-hards or fanatics, likewise kept their feet, firm in the stand that where sincerity and logic ruled, no concessions were possible.

This new ground in theological controversy was well plowed in the contemporary novel. In the 'eighties when it was believed the storm had blown over, and a new spirit of concession was in the air, the novel breathed hope. Erring sons who had wandered from the path of faith were shown that they had no need to stray. But in the decadent 'nineties and even in the next decade, the incompatibility of the ape and the angel theories once more was enunciated, and ridicule heaped on those who deemed they could be mated. Consideration of this phase of the relations of religion and science in the novel will turn on the attitude of the novelists toward the problem of reconciliation. Distinction will be made between those novels which faced it with sanguine expectation, and those which regarded an honest reconciliation as neither reasonable nor practicable.

I

The few novels that hold out hope for reconciliation suggest no actual concessions, but merely an acceptance of as much of the doctrine of evolution or Darwinism as may prove compatible with the belief in a beneficent God. A new teleology is fashioned to take the place of the Paleyan one destroyed by the doctrine of evolution, and a retranslation of the Bible serves to resolve all the cruxes which the new theories created. However much of evolution such writers were ready to admit, they could not, it seems, bring themselves to jettison the man of Genesis for the man-ape.

As the one churchman eminent in his day who accepted the findings of science, Charles Kingsley has a niche all to himself in the Hall of Truth-seekers. Kingsley was both naturalist and clergyman, but he profoundly believed that the Bible and science were not at variance. Darwin's *Origin of Species* and his book on the *Fertilisation of Orchids* had opened a new world to him, but it made all that he saw around him, if possible, even more full of divine significance than before. Through all his searching he sought and found the hand of God in the work of nature.[3]

Having seen the light himself, Kingsley endeavored to show others how a free mind could accept scientific truth without disloyalty to the truth with which his soul had long been nourished. In a lecture on "The Theology of the Future" delivered at Sion College in January, 1871, he urged upon the clergy of his day the sanctity of scientific investigation.

> I entreat you (he told them), if you wish to see how little the new theory, that species may have been gradually created by variation, natural selection, and so forth, interferes with the old theory of design, contrivance, and adaptation, nay, with the fullest admission of benevolent final causes—I entreat you, I say to study Darwin's "Fertilization of Orchids."[4]

He had no patience with those who challenged Darwinism without knowing "the facts";[5] or with those timid scientists who dared not follow the skein to the light.

Darwin is conquering everywhere, and rushing in like a flood, by the mere force of truth and fact (he wrote the Reverend F. D. Maurice in 1861). The one or two who hold out are forced to try all sorts of subterfuges as to fact, or else by evoking the *odium theologicum*. . . .

But they find that now they have got rid of an interfering God—a master-magician, as I call it—they have to choose between the absolute empire of accident, and a living immanent, ever-working God.[6]

He himself had no doubts that the latter alternative was the true explanation; and "by the strange light of Huxley, Darwin, and Lyell" worked out his conception of Natural Theology. Divinity is indeed revealed to man in the things he can touch and see, but there is a much wider region which we can know only through contact of spirit with spirit. An enormous amount of the world's work is done by causes strictly material, science has shown, but the tendency of physical science is not towards the omnipotence of Matter, but to the omnipotence of Spirit. "And I am inclined to regard the development of an ovum according to kind as the result of a strictly immaterial and spiritual agency."[7]

Kingsley reverenced Nature, then, but he reverenced more the will that is above Nature. His reverence for Nature was not antagonistic, but paid homage to his faith in the supernatural. "We knew of old that God was so wise that He could make all things; but behold, He is so much wiser than even that, that He can make all things make themselves."[8]

It is this last conception—the essential characteristic of living matter to reproduce itself—which informs the deeper philosophy of Nature in Kingsley's immortal *Water Babies* (1863).[9] When the child comes to Mother Nature, and expecting to find her very busy, is surprised at her folded hands, he receives the following wise reply to his natural question: "I am not going to trouble myself to make things. . . . I sit here and make them make themselves."[10]

By keeping his faith in God and the soul Kingsley unflinchingly faced the truths that the natural sciences were unearthing, and himself assisted in the process. That 'trailing clouds of glory do we come, from God who is our home,' he never doubted, and urged others never to doubt: "For then the great fairy Science, who is likely to be queen of all fairies for many a year to come, can only do you good, and never do you harm."[11]

For those who were content in that age to accept the premise of a First Cause, Kingsley's ready resolution of the religious-scientific issue to the effect that God had made things to make themselves, would seem a wholly adequate one. Kingsley was far ahead of his time. Not for many years did another fictional work appear which hewed to the line of the clergyman-novelist's reasoning. *Stronbuy* (1881),[12] by Sir James C. Lees, follows Kingley's suggestive lead. The heroes of *Stronbuy* are addicted to heavy conversations on science, religion, and political economy. In one of these the discussion turns on the indictment of evolutionists as atheists, and the claim of one minister, McKay by name, that "if he had Darwin there . . . he would knock over his fine reasonings in five minutes."[13] But Farquhar, one of the company, cannot see matters in this light. With no doubt that God had set the first impulse of creation going, he compares the Kingsleyan view of evolution of the cosmos developing itself at need to a Watt's steam engine with the miraculous power to develop all necessary improvements. And yet, he concludes, "to hear M'Kay and men of that kind talk, you would imagine people holding opinions like these were Atheists."[14]

There is none of this Scotch sobriety about the mystical thinking of Laurence Oliphant. Anticipating by a few years Marie Corelli's queer concoctions of misunderstood philosophy and science,* Oliphant presents in *Altiora Peto* (1883)[15] an adaptation of the doctrine of evolution to the field of morals. Keith Hetherington, the hero, recognizing the part the process of evolution has played in the physical and mental development of man in the past, puts his faith in the thesis that the same process is going on to develop man morally. He sets on foot a religious movement to reconstruct society, to inaugurate "an evolutionary period . . . in the inner man . . . and make him a far more fit inhabitant of the world." This is Keith's idea of how the survival of the fittest is to be brought about.[16] As his chief disciple he enlists the heroine, Altiora, who under the master's teaching detects in herself a "protoplasmic modification." Together, to her death, they attempt to effect a regeneration of moral life on an altruistic basis "by persons undergoing an evolutionary process."[17]

Kingsley, Lees, and Oliphant wrote in perfect assurance that religion and science would make peace. But that was before *Robert Elsmere*. After the publication of Mrs. Ward's novel, polemical treatises in the form of fiction made every man in the street conscious that the fight for the soul of man was still going on. For what one novelist might see as an insurmountable obstacle dividing Genesis and geology, another novelist might characterize as merely a mote in the eye of the first writer, and proceed to bridge the gap with consummate ease. Such a *tour de force,* wedding the ape and the angel, is accomplished in *Paul Nugent, Materialist* (1890). With the too evident purpose of refuting *Robert Elsmere,* the authors of this novel set up an atheist dummy who with charitable compliance puts out his chin and asks to be knocked down. The glib arguments of the senior curate of Elmsfield, the Reverend Herbert Lovel, may have shown Nugent the error of his ways;

* Oliphant was later (1888), the author of a book, *Scientific Religion; or Evolutionary Forces Now Active in Man.*

but the declaration by the beauteous saint to whom the materialist has proposed marriage that she "would rather die than marry a man who denied her God"[18] may—of course quite unconsciously—have facilitated the work of conversion.

As the son of an evolutionist who, an avowed atheist, "out-darwined Darwin," the hero of the novel, Sir Paul Nugent, fell heir to all his father's materialistic philosophy. Friends and relatives, all without faith, confirmed him in his inheritance. Evolution was his religion. So circumstanced, he has the misfortune to fall in love with Maude Dashwood whose Christianity was not only orthodox but aggressive. Maude shudders with disgust at his irreligion, but becomes inspired with the mission of bringing him within the fold. One tearful look from her and Sir Paul "saturated with the teachings of Huxley, Tyndall and Spencer"[19] is ready to give up the fruits of years of thought and research.

With the Reverend Herbert Lovel, Nugent discusses Robert Elsmere's desertion of his faith. With ease the minister counters every argument which destroyed Elsmere's belief, and new ones which Nugent half-heartedly suggests. The materialist's resistance, always weak, becomes feeble. It begins to fade altogether when Maude dangles her bait before him—she will not marry an atheist.

Victim of an accident, Paul is confined to his bed. Lovel visits him. Their conversation turns inevitably to the conflict of the Bible and evolution. With lordly condescension Lovel rehearses the arguments by which the Bible was made to swallow scientific fact. The six-day creation, he instructs Paul, has been made elastic enough to encompass the aeons of time required by evolution, for the Hebrew word "yom" previously misinterpreted as "day" has now been properly rendered as "age"; the Noachian deluge did not take place over the "earth" but over a "region"—probably, western Asia; and the crux created by the

Genesis account of light in the universe before the creation of the sun and moon, has been solved by the theory of luminous nebulæ. Since science "was not even in its babyhood when this story of creation was written, therefore it must have come to the writer by direct revelation from Heaven."[20]

Lovel acknowledges his debt to Kingsley for the explanation which makes evolution a manifestation merely of God's infinite wisdom in relieving himself of the business of making separate creations. But he dismisses airily the theory of man's descent from an ape-like ancestor: there are too many missing links. Science, he pontificates, will never bridge the gap between the brain capacity of the highest ape and the lowest human.

"Compare your intellect with that of the cleverest monkey you ever saw," the Reverend Lovel suggests.

Paul made a grimace. "I certainly feel an immeasurable distance between myself and the brutes."[21]

The clergyman presses home the argument for a First Cause, and Paul avows himself completely confounded. Finally, in the chapter "Materialism Conquered," the abashed atheist embraces a faith and a wife. His friends would have it that this change was due "to the emasculated state of his intellect" during the severe illness; but Sir Paul maintains stoutly that "his convictions were the result of his researches."[22]

II

The series of novels which opposed the attempt to syncretize Scriptures and science carried on their campaign principally in the closing decade of the nineteenth and the opening decade of the twentieth century. Just as Charles Kingsley anticipated by many years the attempt to establish an *entente cordiale* between the factions, so William Hurrell Mallock anticipated the trend away from reconciliation. In the work of Kingsley, Lees, and Oliphant there was evident the conviction that from the flaming pyre on which evolutionary doctrine had placed religion would emerge a new phoenix more glorious than the old. In the work of Mallock and his successors there is manifest the belief that not even the ashes of the old supernatural faith can be utilized in the construction of a new. And only ridicule is visited on the conception that the fact of science and the faith of religion have anything in common.

No amount of wrenching, it was felt, could make the square peg of evolution fit into the round hole of religious dogma. Various points of difference are italicized in these novels to illustrate this conviction. That in man's descent from the ape stock there is implicit a denial of man's divinity which no form of Christianity can admit, is one important difficulty stressed. Further, it is alleged, evolution and Darwinism not only deny design, but provide all the machinery necessary for the evolution of those elements of mind and morality which we call soul. The attempt to reconcile Genesis and the Darwinian doctrine struck most as the crowning absurdity. Such an attempt, if seriously made, would make a mockery of religion, it is suggested. The evidence of geology cries out against the cosmogony of Genesis in a voice that will not be stilled. Other novels, assuming the possibility of such reconciliation in another age than this, insist that neither the churchgoer nor the Church is prepared to scrap St. Paul for Darwin, the saints for Huxley.

Son and grandson of clergymen, cradled in an atmosphere of strict conservative tradition, William H. Mallock felt no call to deny either his heredity or his environment. Adolescence and early manhood found him in a ferment over the menace of orthodox religion: the result it seems not so much of a deep religious faith, as of a fear of the tragic eventuality should supernatural faith be eliminated.

He set himself the project of discovering the absurdities of current liberal philosophy and of finding for himself a foundation for some positive philosophy of life fortified by reason.[23] His views he propounded not alone in prose theses but in novels also to gain a wider public. Though something like the germ of the religious novel had a pre-existence in earlier works by other authors, to Mallock perhaps belongs the credit of being its pioneer. It was not till Mrs. Ward had given to the world *Robert Elsmere,* however, that an appetite was developed for this kind of literature.

By the application of the method of a *reductio ad absurdum,* Mallock sought to show the ridiculous side of several distinct schools of thought in *The New Republic* (1877).[24] Especially did he direct his irony against various species of unfaith, and even against attempts to reconcile the facts of evolution with the belief in God and design. The work is a sort of Peacockian revival of personalities in which the characters are easily recognizable as prominent contemporaries. In the featured conflict of science and religion, Professors Huxley and Tyndall are represented by Mr. Storks, the prosaic, and Mr. Stockton, the sentimental materialist, as protagonists for the one side; and John Ruskin is personified by Mr. Herbert for the other. Dr. Jenkinson (Professor Jowett) is the great Broad Church divine who tries to resolve the traditional and the scientific view of the world and man.

The New Republic is a series of dialogues in the manner of Plato set in the framework of a summer week-end at the sea-side. Otho Laurence invites a select circle of friends to spend a "Saturday to Monday" at his marine villa. In order to avoid the uneasy silences and unnatural conversations at meal times he has prepared a novel scheme. Just as the chef has drawn up a menu for the meal, Mr. Laurence has devised a menu of conversation to which all have to adhere.

A series of skirmishes ensues in which Mr. Storks "who is great on the physical basis of life and the imaginative basis of God" breaks lances chiefly with Mr. Herbert; Mr. Storks insisting strongly on the part evolution has played in the progress of species, and Mr. Herbert expressing as arrogantly, "Had I only the power I would myself put a forcible stop to all this evolution."[25] But the high spot of the pastiche is Dr. Jenkinson's sermon in the private chapel on Sunday, developing his theory of "moral evolution." That man has been produced by the "process of evolution from a beastly and irrational ape" Dr. Jenkinson does not doubt; nor would he have any others doubt it merely because it is repugnant to their sense of dignity. Man is not less truly man, Dr. Jenkinson points out, if he has sprung from an ape. Modern science, moreover, should hold no terrors for the Christian, for the principle of evolution can likewise be applied to the history of Christianity. Just as that theory shattered belief in an immediate creation of all species, so it has shattered the concept that Christianity came into the world at once, a ready-made body of theological doctrine. Through long struggles for its existence "Christian perfection emerged gradually out of imperfection."[26]

To Mallock the supreme absurdity of the Broad Church liberalism was that it went so far as to embrace in its fold and to label "true Christianity," all opinions—even such, as the theory of evolution, which constituted an outright denial of Christianity.

Once again, in *The Veil of the Temple* (1904),[27] written more than a quarter century later, Mallock strove to show the absurdity of all current attempts on the part of clerical apologists to vindicate religious belief by scientific reason. The method used is that of *The New Republic,* a series of debates on set questions in which the logical consequences of religious and scientific doctrines are pushed to extremes.

Chief protagonists are Rupert Glanville, a writer on scientific philosophy, and his friend Alistair Seaton, a Hegelian professor who holds for religion. Glanville is a determined evolutionist and sceptic with a private museum in which the evolution of man is graphically represented over a span of a million years. Glanville's guests at his Irish estate visit with him a clerical Congress where are being held conferences on religion and science: on "The truth of Genesis attested by evolutionary Science," "The Downfall of Darwinism," and others. "They're going to prove Genesis by Darwin and knock down Darwin with Genesis," one of the clergymen tells Glanville.[28] These conferences of the orthodox suggest to one of Glanville's guests the idea of holding a conference of the unorthodox, on the conflict of the Bible and science.

In succession the theological arguments usually advanced either to refute science or to reconcile faith with the fact of science, are scrutinized and discarded. Bishop Wilberforce who had called the doctrine of evolution "the frenzied inspiration of the inhaler of a mephitic gas" and denounced Darwin for representing famine, disease, and death generally as the means by which nature has selected the creatures fittest to live, is sneered into disrepute. Using the Darwinian hypothesis as criterion Glanville demonstrates the indifference of nature to the individual, "which fact lies at the very root of organic evolution," yet is a razor applied to the throat of natural theism, the root idea of which is that the individual is nature's hero.[29] Professor Seaton's suggestion that possibly God's designing hand is to be seen in the production of varieties on which natural selection operates is confuted by Glanville's quotation from a Mr. Cosmo Brock (Herbert Spencer). Brock stigmatizes the argument for God as the Great Contriver by pointing out that for wanton recklessness and stupid fumbling, nature's method of producing varieties has not been surpassed: in order that one creature may be produced fit to survive, a multitude are produced fit only to perish miserably. The theological view that would hold man a thing apart from other species Darwin has destroyed completely, Glanville insists. The human organism has been connected not only with the monkey, the dog, and the fish, but with the simple cell, and through the cell with the vegetable world.

"Do you mean," said Mrs. Vernon, "that we are now asked to regard ourselves not only as the children of monkeys, but as the grandchildren of beans and potatoes?"[30]

And finally, by elaborating on the scientific evidence that establishes conscience and consciousness as subject to the laws of evolution, Glanville throws overboard the *apologia* that though the body may have evolved, it is the mere envelope of the soul implanted by God.

Professor Seaton, who has yielded step by step, at length accepts the whole of the evolutionary gospel. But none of those present is ready to accept scientific negation completely as a substitute for the gospel of Faith; only the latter, they feel, can be a guide to action. Glanville gives his personal resolution of the problem by urging on all the pursuit of Truth as a religion.

Frederick Hazzleden (1887)[31] by Hugh Westbury is a story of politics and Fenian conspirators relieved quite frequently by the letters and conversation of old Mr. Hazzleden, an amateur horticulturist who is bitten with Darwinism to such a degree that he can never refrain from mounting his hobby no matter in what company. He attempts to communicate his evolutionary enthusiasms to a shoemaker, to the curate, to his sister Maria, with no success on any of these fronts. The bootman didn't know much about Darwin, "but he knew a man who had a mill there."[32] To the curate the Darwinian lent a copy of the *Origin of Species* but with small hope that it would do him much good. The clergyman had previously failed to respond to old Hazzleden's attempt to demonstrate consciousness in plants and to prove a link between animal and vegetable life. He merely

. . . screwed up his eyes and said something about the beautiful designs of Providence. All bosh, Fred, and I told him so. That vine has what Darwin calls a low form of consciousness as yours and mine, only less of it. . . . For my part I should not be the least ashamed to trace my ancestry to a cabbage.[33]

His constant preoccupation, when not applying to his plants the Darwinian principles, is trying to convince Maria of the truths of evolution, by lecturing her or by reading to her from the "good book." But Maria has a turn for theology and finds a spirit akin in her neighbor Mrs. Williamson. Between them they have decided that since it is contrary to the word of God, the evolution theory "will no' do." Their sense of dignity is assailed by the notion that patriarchs and apostles, as well as themselves, are descendants "from the hugly reptiles as orgin-grinders carries about with them." Both try to bait Mr. Hazzleden out of his convictions. "It's an insult to us," said Mrs. Williamson. "My family was most respectable people, and never 'ad no relations with hapes or any other disgustin' beasts."[34] Maria, with better control of her aspirates, but equally ill logic, echoes, "By your irreligious theories you would compel us to believe that the Being whom we worship was descended from the beasts. . . . Is it not written, 'Man was made upright, but they have sought out many inventions'?"[35] But Mr. Hazzleden, an open volume of Darwin upon his knee, was "meditating on the origin of species"—fast asleep.

Mr. Hazzleden's son Frederick stands for Parliament as representative of Dockborough. This leads to another gratuitous intrusion of the problem of conciliating the forces of science, a problem that faces the fashionable Broad Church chapel of Mostyn Mount in Dockborough.

The Mostyn Mount parishioners were receptive to the ideas of Darwin, Tyndall, and Huxley, but confronted by the alternative of giving up Adam and Eve and receiving Darwin, they hesitated, then debated. One venerable deacon held that the trouble came from reading secular literature. During his life he had read nothing but the Bible, hence had not the shadow of a doubt of the Mosaic account. Another deacon had read a few chapters of Darwin and "did not see that there was much in them."[36] But the Reverend Samuel Robinson, the young pastor, a man of fine literary taste, proclaimed his inspired resolution of the difficulty, and ended immediately the disputes of the deacon and the depression of the Mostyn Mounters. Genesis was a poem, a Hebrew *Paradise Lost!* Never more was the Mostyn Mount chapel perturbed by difficulties connected with the Mosaic cosmogony.

A more reasonable settlement of the difficulty than was offered to the Mostyn Mounters, Olive Benison, heroine of Miss Theo Gift's (Mrs. Theodora Boulger) *Dishonoured* (1890?),[37] offered to representatives of Church and of Science. Neither would accept her suggestion.

By lending a quarryman a copy of Hugh Miller's *Testimony of the Rocks* (a pre-Darwinian (1857) geological text questioning the seven-day creation), the heroine provokes the ire of the narrow-minded Presbyterian, Reverend Campbell. "Do ye no ken that this buik is an eenfidel one and written to controvert the Buik o' buiks an' the Wurrd o' God!" he tells Olive.[38] But for all his fulminations she refuses to accede to his demand that she throw the book in the fire. Stoutly she defends the thesis that we cannot love God the less for knowing more about the wonders He has created for our benefit. But the old minister will have none of this compromise.

Resolved to undo the evil she may have done in lending the quarryman so questionable a book (she herself has read only a part), Olive prepares to go the length of sacrificing "Hugh Miller on the altar of Moses and of the Pentateuch."[39] But the quarryman puts up unexpected resistance. He will not permit the sacrifice. On the contrary, with a strange wisdom he proceeds to sacrifice Pentateuch on the altar of geology. Professing a belief in God and in Miller's religiousness, the quarryman, nevertheless, offers the rocks about them and fossil incrustations in them as evidence of the aeons of time necessary for species to evolve. "Those who want to 'reconcile' Moses and geology . . . have set themselves a work as puerile as it is unnecessary," he exclaims, denying Revelation.[40] But his eloquence gives him away: he is not the ignorant quarryman she has supposed, but Stephen Ralston, F. R. S., authority on the fossil cryptograms of the West Indies.

George Gissing, who in *Workers in the Dawn* (1880)[41] had presented a treatment of lost faith, utilized again as a theme the impact of the Darwinian discoveries on a believing world in his novel *Born in Exile* (1892).[42] Here as elsewhere Gissing depicted the struggle of one who was not armed for the conflict of life, but who strove anyway, and in the end fell without glory.

Young Godwin Peak, an agnostic, was born into the wrong family, into the wrong social class—born in exile. The advantages of higher education gave him social ambitions. Impelled by his love for the highborn Sidwell Warricombe, he guilefully professed a Christianity he did not feel. Detected in the fraud, he dissipated the rest of his life, broken on the wheel of circumstance.

An astute thinker and evolutionary scientist, Peak sounds off belligerently on the current trend toward harmonizing Darwinism and the Bible. With dramatic irony Gissing has his hero damn such middle-of-the-road tactics, as well as the timorousness of agnostics without the courage to air their convictions. Nor does Godwin Peak refrain from voicing his contempt for "those solemn jackasses who brayed against Darwin" ten or twenty years before. Full of this spleen, he composes a brilliant satire at the expense of those quasi-scientific writers who affect to make peace between the arch opponents, evolution and the Bible. It is published in the *Critical* review anonymously, and is acknowledged everywhere. But in the meantime the wind has changed: Peak has fallen in love with Sidwell, daughter of the geologist Martin Warricombe. Filled with the determination to climb socially, the confirmed agnostic decides to study for the ministry to secure a firm social position, even though it would mean playing hob with all his cherished scientific ideals.

Shrewdly, he learns to make use of Martin Warricombe as a crutch. The old man, for all his scientific leanings, yearns for the triumph of orthodoxy, or else to see certain and sure removal of the inconsistencies between the tenets of his faith and those of science. With venial hypocrisy, Godwin Peak essays to establish the truth of religion to Warricombe's satisfaction. At the same time he is made doubly conscious of the futility of this line of reasoning by a chance meeting with the Reverend Bruno Chilvers, his former classmate, now a brilliant success as a latitudinarian minister. Broadest of the Broad Church apologists, Chilvers astounds Peak with his doctrines. The minister is all for the infusion of the scientific spirit into the work of the Church. No compromises will satisfy him; on the basis of scientific revelation he would construct his spiritual edifice. "Less of St. Paul and more of Darwin! Less of Luther, and more of Herbert Spencer," will be his shibboleth, he announces to Peak; and he would begin by preaching a series of sermons on the *Origin of Species*.[43] But Chilvers realizes too well that the day when his parishioners will be equal to receiving such heterodoxy is far, far off. Soon afterward, the death blow is dealt Godwin Peak's soaring hopes. The unearthing by Sidwell's brother of the satire Peak wrote for the *Critical* in his agnostic days, betrays the ambitious young man. His romance and life tragically crumble to ruins.

By his awareness that the Anglican Church even at its "broadest" would find certain Darwinian doctrines not palatable, the Reverend Bruno Chilvers saved himself from wrecking his career. By his failure to anticipate that the Catholic Church would refuse sanction to the marriage of Darwin and dogma and drive them apart, Father Benecke in Mrs. Humphry Ward's *Eleanor* (1900)[44] owed his personal tragedy.

Whatever may be the appeal of the pathetic romance of Eleanor, Lucy, and Manisty in Mrs. Ward's *Eleanor,* the reader will come away from the novel with more than a little sympathy for the plight of Father Benecke. With deftness Mrs. Ward has woven into the main plot the mental storm and stress of this excommunicated Bavarian priest, guilty according to the Roman church of having wished to reconcile Darwin and biology with Catholic dogma.

His book, "a little Darwinian yeast leavening a lump of theology,"[45] is damned as heretical and placed on the *Index,* and forbidden to Catholics. Father Benecke must submit—or be excommunicated. He weakens, and on the promise that no public shame will attend his recantation despatches his letter of submission. But the Church plays him false and publishes his letter. Shamed to the quick, filled with remorse at betraying the truth, Father Benecke repudiates his submission. Instantly the whole weight of the Church's displeasure falls on him. He is suspended from his priestly and academical functions (he was a professor of theology), and forbidden the sacraments. All Rome treats him as one dead.

Hounded by the Jesuits, he is driven to seek refuge in a secluded mountain town, Torre Amiata. But even there the stigma follows him. The Bishop passes the word: *E un prete spretato* (He is an unfrocked priest), and ostracism ensues. No one will serve him, or offer him shelter, whatever the payment. The clergy pass him, and turn away their heads; the peasants fear him as a curse on their crops; the children, with faces shining with malice, cry out at him, *"Bestia!—bestia!"* stone him, and flee. But the old man—he was sixty-five—bears his cross stoically. Only once does he beweep his outcast state. "Excommunicated. And for what—because I said what every educated man in Europe knows to be true?"[46]

His talks with the intellectual Eleanor who has isolated herself at Torre Amiata rescue him from his tortured thoughts. The rationalist in him is stimulated into renewed activity, and Father Benecke experiences a new birth of freedom. He has been brooding over a revision of his book. He makes an instantaneous decision. No longer will he concern himself "with recapitulating within Catholic limits the vast accumulation of biological fact and historical criticism."[47] No longer will he defer to authority. His book, instead of being revised, must be rewritten. He will rewrite it, not as before, a rationalization and compromise with Church doctrine, but free, untrammeled in its espousal of truth.*

Reconciliation which the Catholic Church refused to admit in *Eleanor,* the scientific world and clerical world both declined to admit in Edmund Gosse's *Father and Son* (1907).[48] Differing in kind, the tragedies that resulted in the lives of Mrs. Ward's Father Benecke and Gosse's Father, were alike profound.

---

* What that truth was, Mrs. Ward had, of course, already shown in *Robert Elsmere.* She reiterated it in *The Testing of Diana Mallory* (1908): "What faith was to the thirteenth century," Ferrier, a political leader in that novel, tells the heroine, "knowledge is to us. St. Francis rekindled the heart of Europe, Darwin has transformed the main conception of the human mind."

Mrs. Humphry Ward, *The Testing of Diana Mallory* (New York, 1908), p. 344.

In spite of his great and voluminous achievements as a critic and a recorder, Edmund Gosse will chiefly live—will permanently live—as author of *Father and Son.* Published in 1907, the work is described in the opening lines as "the record of a struggle between two temperaments, two consciences, and almost two epochs." The form is that of the novel, but it is truth which Gosse tells—the biography of his father, an autobiography of himself.

His father, in the novel, emerges as a figure of iron-strong faith, devoting his life to the passionate pursuit of religion and science in their most irreconcilable forms. Gosse does not pretend that the inconsistencies were susceptible of any intellectual reconciliation. But the tragedy of the father's life is that he did believe compromise possible, and split his heart on this rock of delusion.

For the Father religion existed in William James's words "not as a dull habit but as an acute fever rather," a mania bordering on the insane. His profession was that of zoologist and writer of books on natural history. With the same single-minded integrity and lack of imagination that set him laboring at a literal interpretation of the Bible, this "honest hodman of science" methodically collected facts and marshalled observations. Not for a moment, however, did he permit his scientific studies to interfere with the religious. Indeed, the more science threatened to encroach on his peculiar religious beliefs, the more fiercely were those beliefs proclaimed and their exercise observed. But there came a time when the two could not be kept apart.

It was the year 1857, "the great moment in the history of thought when the theory of the mutability of species was preparing to throw a flood of light upon all departments of human speculation and action."[49] Lyell, Hooker, Wallace, Asa Gray, Wollaston, each in his own sphere, were coming closer and closer to a perception of that secret which was first to reveal itself clearly to the genius of Darwin. Indeed, in the year before, in 1856, under pressure from Lyell, Darwin had begun that "abstract of an essay" which developed so mightily into the *Origin of Species*.

Word had gone out of the nature of this new discovery in this period of intellectual ferment to a selected group of naturalists—and of the chosen, the Father was one. His opinion was solicited. A choice had to be made.

The facts stared him in the face. "With great and ever greater distinctness his investigations had shown him that in all departments of organic nature there are visible the evidences of slow modification of forms."[50] But what about Genesis? Geology certainly *seemed* to be true, but the Bible, which was God's word, *was* true. The belief in a direct creative act from without, peopling the world with a sudden, full-blown efflorescence of fauna and flora, was a part of his very being, and he would have abandoned the entire study of science sooner than relinquish it.

Inspiration came. He would reconcile the phenomena of geology with the Mosaic record, even an exact and inelastic interpretation of that record. For one believing implicitly as he did in the fact of creation his compromise was logical, inevitable. It was this: When the catastrophic act of creation took place, the world immediately assumed its present stratified structure. There had been no development.

He called his work *Omphalos*.

Raucous was the laughter with which the work was greeted. It was the notion that the world was created with fossil skeletons in its crust which met with the most ridicule. The author was charged with supposing that God had formed these objects on purpose to deceive—in order, in fact, to set a trap for naughty geologists. Scornful silence was all the book received from the scientists; and the orthodox party whom he had expected would rally round him, laughed and threw it away. Kingsley wrote him:

> I cannot believe this of a God of truth. I cannot give up the painful and slow conclusions of five and twenty years' study of geology and believe that God has written on the rocks one enormous and superfluous lie.[51]

So bitter a reception was not without its effect: Gloom chilled to anger, anger not alone with his fellow mortals but with God for so evilly rewarding his efforts. But soon conscience made its voice heard, and anger turned to self-accusation. God had brought about the failure of his attempt at the reconciliation of science with religion as a punishment for sin. Rigorous was the self-examination, rigid the ascetic spirit; and in penance he laid upon his harmless pleasures taboo after taboo. Sensitive now to his own alienation from grace, his captious spirit set to work on his son. Eventually their natures clashed, rebellion broke out. Inevitably, as in *Christopher Kirkland,* the schism came: the two generations, thesis and antithesis, were driven apart.

> No question is more often put to me regarding my father than this (wrote Edmund Gosse): How did he reconcile his religious with his scientific views? . . . The word 'reconcile' is scarcely the right one, because the idea of reconciliation was hardly entertained by my father. He had no notion of striking a happy mean between his impressions of nature and his convictions of religion. If the former offered any opposition they were swept away. The rising tide is reconciled in the same fashion to a child's battlements of sand along the shore. It was under the pressure of this sense of awe that, when his intellect was still fresh, he deliberately refused to give a proper examination to the theory of evolution which his own experiments and observations had helped to supply with arguments.[52]

That a literal application of the philosophy of evolution to religion would consume that religion just as surely as a torch applied to straw, was the ultimate conviction of those novelists whose thoughtful treatment of the question we have just considered. None of them, however, is over-zealous to conduct the Deity to the verge and, in the language of Comte, "bow Him out with thanks for His provisional services." Rather there is patent a worship of truth so strong that the authors would have it prevail at whatever cost or sacrifice: the immolation of a spurious God, if necessary. Dr. Buckland, a prominent and energetic scientist of the clerical order, had once uncompromisingly asserted that all scientific teaching must be forever subordinated to the cosmogony of Genesis. Just as uncompromisingly these novelists insisted that the cosmogony of Genesis and all other aspects of faith must be forever subordinated to the teaching of truth. Thus, despite his orthodox training, William H. Mallock was willing in *The Veil of the Temple* to subject every tenet of his faith to the scrutiny of science, and to be satisfied with the irreducible minimum of religion: "Keep the soul alive by following truth in one form or another."[53] That the teaching of truth could not be carried on with impunity was the theme of a number of novels. Olive Benison of *Dishonoured* drew on herself the rancor of the ministry by daring to educate a quarryman. The Reverend Bruno Chilvers in *Born in Exile* dared not preach from the pulpit those scientific doctrines which he had taken to his own bosom. And in *Eleanor,* Catholicism, refusing to accept Father Benecke's compromises, drove him out of the Church to face the slings and arrows of outrageous fortune. All three protagonists believed firmly in the truth they were espousing.

The danger of straddling two horses is always that of falling between them. The attempt of the Mostyn Mounters in *Frederick Hazzleden* to achieve a liberal resolution of the Genesis and geology question by identifying Genesis as a poem was manifestly ridiculous. Godwin Peak in *Born in Exile* fell trying to maintain a reconciliation he knew was specious. In *Father and Son,* the Father's hapless compromise-idea was trampled under foot by scientists and theologians alike, smashing the man's self respect. Certainly, as in the case of Father Benecke and the elder Gosse, the futile hope of reconciling evolution and religion made for as tragic a theme as the loss of hope of salvation. To the Victorians wandering between two worlds these were the problems out of which *Hamlets* were fashioned. Today we view them with cold detachment. The religious issue, burning hot in those antimacassar days, is now but a flaky ash beneath the live coals and roaring flames of class struggle, fascism, and war.

Sir Charles Lyell quotes a saying of Professor Agassiz "that whenever a new and startling fact is brought to light in science, people first say 'it is not true,' then, 'that it is contrary to religion,' and lastly, 'that every one knew it before.' " The controversy over the truth of the Darwinian hypothesis, however, arose principally from the fact that theologians persisted in affirming that "it was not true" because "it was contrary to religion." But soon it was being complacently remarked that not alone evolution but natural selection also were of ancient date, "as old as Lucretius," and concerted attempts by the Church to bait the apes with the angels came to an end.

The sudden cessation of opposition to the Darwinian theory in the last decades of the nineteenth century, and the sudden cordial acceptance of the doctrine by the secular masses and by the Protestant Church, are among the curiosities of psychological history. The surprising, unconditional surrender had wide ramifications. In the field of literature in particular it brought about directly the collapse of a rich dramatic theme, the theme of spiritual conflict. The *Sturm und Drang* of a mind forced to choose between the traditional revelation of God and the radical revelation of evolution, had offered a happy hunting ground for the psychological-minded novelist, coinciding significantly as it did with the era of the psychological novel. But once, as we have seen, the Church ceased its childish attempts at theological apologetics and adapted its beliefs to the new theories, the element of dramatic conflict disappeared. Science and religion went their respective ways; the interest of the public once at fever-heat, cooled, then evaporated altogether; and the theme practically disappeared from the pages of the novel.

CHAPTER NINE

The Anthropological Romance

During the last half century a number of significant discoveries have been made in regard to the fossil races of men and near-men. These ancient types constitute six distinct links, which are no longer "missing," between modern man and anthropoids. It is not certain that any one of them is a direct ancestor of man; they may be simply collateral branches of our ancestral tree. All these extinct species and genera of man, as we know by the age of the rocks in which their bones were found, have been associated with the Pleistocene Age not more than 500,000 years ago. A possible exception is the fossil "Dawn" man found at Piltdown in Sussex who is believed by some to date back to the Upper Pliocene, the beginning of the close of the Age of Mammals. Prevalent opinion among palæontologists is that the *Hominidae* separated from the *Simiidae* not less than one or two million years ago. Thus the Age of Man, it is assumed, dating from the time when man began to use fire and make tools, has endured for the last 1,250,000 years. This immensely long period is only about one-one thousandth part of the time which has elapsed since life first appeared on the earth.

Palæontologists generally suppose that the modern species, *Homo sapiens,* appeared near the end of the last or fourth Ice Age probably 20,000 years ago; the Neanderthal* species (whose discovery in 1857 in the Neanderthal near Düsseldorf, Huxley treats at length in *Man's Place in Nature*), before the last Ice Age, from 50,000 to 100,000 years ago; the Heidelberg species (found in 1907) during the second interglacial period, at least 300,000 years ago; the Piltdown species during the first interglacial period,

---

* It has been the custom to name the races of early man after the locality where his remains, or evidences of his presence, were discovered.

say 400,000 years ago; while the Java man (found in 1892) probably preceded the first Ice Age and therefore lived about 500,000 years ago. The Java man, *Pithecanthropus erectus* (erect ape-man), discovered in 1891-2, is the "Missing Link" of journalism, the ape-man of science. Of all the fossil men the Java man is not only the earliest but the most ape-like, the size of his brain roughly midway between that of the modern man and that of the modern ape. The estimated brain capacity, 940 cubic centimeters, approaches that of such low races as the Australian bushmen, and lies between the 1200-1500 c.c. volume of brain in modern man, and the 600 c.c. upper limit of the anthropoid brain. Pithecanthropus was, according to Sir Arthur Keith, "a human being in stature, human in gait, human in all his parts save the brain."

After the human form came into existence, several different kinds of men evolved that no longer exist. They came to the end of their line and stopped, leaving no descendants. The causes of this waning remain very obscure—sometimes environmental, sometimes constitutional, sometimes competitive. Sometimes the introduction of a new parasite, like the malaria organism, may have been to blame. To such a dead-end came the "Cave-Men" or "Men of the Old Stone Age"—the Neanderthal and Cro-Magnon species which lived in Europe during the immensely long period when the great glacial ice-sheets were slowly drawing back northward before the increasing heat of the summer seasons. Neanderthal man was stocky, with ape-like orbital ridges and low brow. He had learned to use his hands as a huntsman and worker of flints, but he had to make way for a superior race and his kind came to an end. Nor did the successors of the primitive Neanderthal, the high-browed intellectual Cro-Magnons, remain for long. Though this magnificent race of warriors, of hunters, of painters, and sculptors, was far superior to any of its predecessors, for some unknown reason it disappeared from Europe, leaving the continent to be repeopled by Asiatics. On the cave walls in the Dordogne region in France, and in the Pyrenees and Northern Spain, the Cro-Magnons left behind evidences of surpassing artistry, frescoes of the woolly mammoth and the woolly rhinoceros they had hunted.

Successively, pre-historic man passed from the Old Stone Age (about 15,000 years ago), then on through a Copper Age, a Bronze Age, and an Iron Age, finally to historic times. Man's struggle upward from the beast is a success story, a story of infinite attempts to adapt himself to nature, of countless defeats, of significant discoveries.

A quarter of a million years ago the utmost man was a savage, scarcely articulate, dwelling in caves, armed with a rough-hewn flint or a pointed stick, naked, living in small family groups, killed by some younger man just as soon as his first virile activity declined. In only a few temperate and sub-tropical river valleys was he to be found, with his little herds, a few females, a child or so.

He knew no future then, no kind of life except the life he led. He fled the cave bear over the rocks full of iron ore and the promise of sword and spear; he froze to death upon a ledge of coal; he drank water muddy with clay that would one day make cups of porcelain; he chewed the ear of wild wheat he had plucked with a dim speculation in his eyes at the birds that soared beyond his reach.[1]

Somewhere along the line, about 75,000 years ago, primitive man learned how to make fire. In caves and in the depths of the earth, intermingled with articles of millennial antiquity, palæontologists have come upon the oldest bits of charred wood and bone. Prometheus had been almost a million years coming.

Fire was but one of many miracles, withal the greatest that happened to primitive man. Man put a finer edge on his wit by making fire; another, when he sharpened one end of a small slim bone, put a hole in the other end, and used it to sew skins together for clothing. By this time, surely, he had gradually attained to hammer and club, axe and spear, borer and flint-saw. The using of stones as weapons or even as implements is occasionally seen among monkeys, but man alone is a tool-*maker*. As he learned to convert skins into clothing and to make use of fire the range of his activities was widened. He could penetrate colder regions. As a hunter he could find more game. Arrow-heads of flint, he soon discovered, could get for him more pounds of meat to the mile. Probably by accident he learned that meat scorched by fire tasted better than raw meat. With fire indeed began man's domination over nature.

We are bound to think vividly of early man as an inventor. Each new invention meant an enrichment of the social heritage. There might be centuries, of course, without a single invention; but there was a cumulative growth. He who first laying hold of a floating tree trunk found it would bear him up in the water, had made a beginning in navigation. And he who first discovered that stones or blocks too weighty to lift by hand could be pried up and moved along with a stout stick, or rolled on two or three round poles, had made the first step in transportation.

We cannot guess in what rude and early age man, observing nature's sowing, began to imitate it and develop the art of civilization. Such beginnings are hidden in obscurity. One of the earliest forms of cultivation, we can surmise, however, was probably that of artificial selection, clearing away useless plants so as to give the valuable ones a better chance. A factor in progress, cultivation gave man greater independence; it made storing possible, as of cereals; and it opened the doors to more people living together in a limited area.

Some time, too, about this epoch the domestication of animals began, of horses, cattle, sheep, goats, and pigs. The Palæolithic Age had passed then into the Neolithic.

age of polished stone implements and weapons. The men of this New Stone Age who came into Europe after the greatest severity of the glacial climate had passed, did not possess the wonderful artistic skill of their predecessors, the Cro-Magnon cave artists. But they made pottery and cloth and erected great sepulchres and temples of stone for their dead. They lived in more or less-settled communities and depended upon an artificial food supply, raising cereals and keeping flocks and herds, instead of depending wholly upon hunting. These, the physical forerunners of modern man, were alike the true forerunners of civilization.

Research in the evolutionary sciences sweeping backward and forward through time and space, enormously broadened men's horizons. Against the few thousand years to which our ancestors had grown accustomed from Bible cosmogony, mankind suddenly found itself dwarfed by the enormous background of time in the past and by the prospect of similar enormous stretches of time for the future. As the new developments in astronomy, geology, and biology began to spread in ever widening circles, interest in science increased proportionately as the rim crept nearer and nearer to man. The Copernican theory of planetary motion, startling in its significance, yet touched only the periphery of man's consciousness: the stars are very remote. But as soon as science began to treat of the ground beneath their feet, of the life pulsing around them, men hearkened to its voice and showed unmistakable interest in the Faustian attempt to

> Get to the root of those secret powers
> That hold together this world of ours.

Concomitant with this widening of perspective and deepening of interest went a rapidly growing demand for books and lectures making the results of science accessible and interesting to the lay public. An enormous body of literature popularizing scientific knowledge sprang up to meet this demand. Veritably the second half of the nineteenth century became a spoon-meat era, an age of universal "boiling down." The solid food of science was reduced to pulp for the general reader to swallow. In some departments the popularizers were the very men whose names stood foremost as original authorities.

The popularization of science—the many opinions to the contrary notwithstanding—by no means extinguished the natural human craving for romance. Science and romance were blended in various proportions to concoct a brew so tasteful that in a short time the scientific romance became an established genre of perennial appeal. Jules Verne's fairy tales of science which had a wide English reading public did yeoman service in preparing the way for the later British product. The species ranges from shilling "shockers" and poorly mixed compounds of passion and science—of blackboard and hymeneal altar, so to speak—to the sternly disciplined and scientifically accurate novel of H. G. Wells. Without doubt such fiction of diverse literary quality has aided the efforts of men like Huxley to make science significant to the popular mind. Most of it was serious in purpose. It sought to interpret imaginatively the implications of science for the future of mankind.

The evolutionary romance embraces the anthropological romance, dealing with the prehistoric past and vestiges of that past in the present; the romance of eccentric evolution, turning on the freakish aberrations of evolution; and the romance of the future. This last division is further supplemented by suggestive hints of what future man may become, presented in a series of novels dealing with the evolution of man-like species in other worlds, other planets.

The anthropological romance projects itself into the past, much in the manner of an historical romance by Scott, and recreates the life and times of prehistoric men and beasts; or setting the scene in the present, represents the phenomenal discovery of living vestiges of the prehistoric.

The romances of the prehistoric past take their cue from the evidences turned up by the pick-and-shovel corps of science. They attempt to make graphic a phase of the development of man before the era of recorded history, in accordance with the discoveries of anthropology and palæontology. Mankind in these romances is usually pictured at some dramatic point in the history of the race, about to turn aside from ape-stock and enter upon his heritage as *homo sapiens;* or on the verge of making a discovery fundamental to his survival and to his ultimate civilization. Two striking illustrations of this specialized fiction are Henry Curwen's *Zit and Xoe* (1887), and H. G. Wells' *A Story of the Stone Age* (1899).

Zit and Xoe[2] is a pastiche, which, had it been written ten or more years later, would inevitably have been labelled a satire on novels dealing with prehistoric human species. It was written, however, in 1887 before such stories became common. Consequently, it must be regarded as merely a whimsical romance of an Adam and Eve just evolved from amid their monkey kindred. The two symbolize the evolution of mankind from ape stock to civilization. Events which in the normal course of human history must have occurred at intervals of hundreds and even thousands of years, are here telescoped into a short span, so that a thousand years is as a day or even an hour.

The parents and brothers and sisters of Zit, the hero, were tree folk. But Zit was "alas! tailless and hairless," and could not walk on all fours like the others, or leap from tree to tree. He was distinctly an anomaly in the monkey family of which he was a member. He could swim and make use of his thumbs, and strut about on his hind legs in a way that was peculiarly exasperating to all. Finally, for the "greatest happiness of the great number," his father bade him depart.

He goes out into a world of mammoths and saurians and his evolution begins at a tremendously accelerated pace. He falls into a heap of sharp, broken flints. These cut him so severely that he is "happily prompted to try if they would not cut something tougher." He then fashions a staff and on it carves the likenesses of the various beasts. Suddenly he comes upon Xoe, coquettish and alone. Naively, Zit submits himself to her will and whimsy and superior culture. She teaches him how to capture and ride a horse, how to make a fire with dry sticks, how to catch trout by tickling them, how to trap deer in camouflaged pits. Newly aware of his nakedness he learns to prepare skins, and to sew them, with cactus spikes for bradawl and tendons for thread, into a garment—a rude tunic. Idyllically, Zit and Xoe live in a hut he has built, sharing discoveries which just about sit up and beg for recognition. He teaches Xoe how to fish with a baited hook made of a splint of bone tied to a tendril. She teaches Zit to make pots. He develops his sharp piece of flint into a formidable flint hatchet, then into a flint-headed spear, and finally into the arrow for a bow. With the same inhuman rapidity and skill Zit discovers how to pilot a tree trunk down the river with a branch. Speedily he improves his sailing facilities by carving a canoe out of a tree trunk, making a broad flat paddle, and then tying half-a-dozen trees together to make a raft. Xoe's scarf serves as sail.

Their home is attacked by hordes of apes (who, strangely without fear, hurl about the remains of their fire, and with human adeptness madly wield hatchets and spears). Zit and Xoe flee to an island where Zit becomes "a husbandman instead of a wanderer, a herdsman instead of a hunter," and domesticates a wild dog which he trains to guard his cattle.

Years pass. His great-great-grandson teaches Zit how to write and thus to write his story.

Zit and Xoe is a motion picture of evolution accelerated beyond belief. Herbert G. Wells's *A Story of the Stone Age* (1899)[4] selects one crucial period in the evolution of the human race which *Zit and Xoe* passed by with the speed of light, and develops it at length. This is the chance invention by a Palæolithic man of a flint-headed axe, a unique advance that intensifies the struggle for life and the survival of the most fit among our progenitors of 50,000 years ago.

Ugh-lomi—Ugh the Thinker—came of a race narrow in the loins and long in the arms, their ears without lobes but with small pointed tips. There was little to mark these human beings off from the wild animals that ranged the country, little except their rudely dressed flints, their poor spears of sharpened wood, and fire—all the weapons they had against hoof and horn, tooth and claw. And there was little at first to set off Ugh-lomi from the rest of the tribe. But jealousy stalked into the squatting-place they called home and the fight for survival began.

Ugh-lomi's first chance to show his mettle came when he pitted himself against the crafty Uya, leader of the tribe. Old but active, Uya coveted the lithe and graceful Eudena. She would have none of him. Supporting her in her rebellion Ugh-lomi drew upon himself the wrath of Uya. Attacked by the whole tribe at Uya's behest, he fought his way manfully to refuge for himself and Eudena. The chance thrust of a stick through the hole of a perforated flint gave him the first axe. With it he accomplished the end of Uya.

Round about Ugh-lomi and Eudena's shelter in exile moved the mammoth, the woolly rhinoceros, the lion, and the cave bear. Successively, Ugh-lomi conceived the idea of riding a horse, and became the first horse-man; used his axe to advantage in the annihilation of a lion, and a club studded with lion's teeth in the annihilation of his tribal enemies; and in the end, with Eudena as his mate, became master of the tribe which Uya had once called his. The fittest had survived. And Ugh-lomi's prowess and superior adaptation to environment had brought the men of the Old Stone Age one step upward in the scale of evolution.

The discovery of "living fossils"—certain types of animals and plants which are evidently the lingering survivors of old world fauna and flora—suggested to some writers of fiction another romantic treatment of evolution. This was to represent the discovery, in the present, of some section of the world where still persist the modified descendants of some ancestral form that lived in the earlier periods of the earth's history.

Probably the first piece of fiction of this kind after Darwin to make use of the theory of evolution was written not by an Englishman but by the Frenchman Jules Verne. His *A Journey to the Center of the Earth* (1864) was probably the literary godfather of most of the English novels which purported to romanticize the doctrine of evolution. Professor von Hardwigg, a scientist, learns that it is possible to go down to the bowels of the earth through an extinct volcano in Iceland. With a nephew to whom he may conveniently explain everything he sees, the professor descends beneath the surface of the earth. They see the geological history of the earth written in the layers of rock by which they pass; and read in them the story of the evolution of plants and animals. A hundred miles below the surface they reach an interior sea on whose banks grow mighty plants of the second period of the world. They catch a fish of the Devonian period, have ringside seats at a battle between an ichthyosaurus and plesiosaurus, and finally catch an awed glimpse of a herd of mastodons guarded by a giant man, the living fossil of some early epoch in the earth's past.

All of the elements of the native English prehistoric

romance which was not to emerge for another twenty years are present in Verne's scientific adventure story. Peculiarly, however, though the tale follows the Vernean practice of making an exacting use of names, dates, and learned discussions of phenomena, nowhere is the name of Darwin or his work even so much as mentioned.

A fertile theme in the English romances was that of the discovery of the "missing link," not in fossil form, but alive. Both *The Quickening of Caliban* (1893) by J. Compton Rickett, and *The Oracle of Baal* (1896) by J. Provand Webster deal with such a discovery.

The Quickening of Caliban[5] has as its hero a big African, Forest Bokrie by name. The scientists who have seen him say "that he is a bit unfinished; not got comfortably through his evolution."[6] Some English showmen, aware of the current interest in science, purchase him and proceed to reap a fortune in London by representing Bokrie as the Missing Link or Ape-man in a series of tableaux, with "plenty of Darwin" in a running comment by a lecturer. They capitalize on the antagonism of the clergy. The advertisements of the Ape-man catch the eye of Christina Ruefold, daughter of an African missionary; of her uncle, Marcus Ruefold, a minister; and of a scientist friend, Professor Racer.

The professor, accepting Bokrie as a genuine living fossil, takes a hand in his development. Bokrie is sent to Cambridge University, but his occasional reversion to bestial type, marked by a wild frenzied dance accompanied by uncouth sounds, "a clicking and barking like a beast of prey," dooms this experiment. The Reverend Ruefold who is Broad Churchman enough to accept the theory of evolution, is fired by a meeting with Bokrie to make a study of the phenomenon for a sermon on evolution. Burning with this message, he refuses to consider his wife's entreaties not to "tear up the Bible in the pulpit." She anticipates the end of his career. But the Reverend Ruefold delivers his sermon without anything disastrous happening. There is a slight ruffling of the usually placid surface of the pews, but no more; and the only repercussion is the chance remark of one parishioner: "Minister seemed extra eloquent last night."[7]

Christina herself undertakes with missionary zeal the conversion of the brute. Her own suspected kinship to Bokrie makes her all the more determined to make a Christian out of him. Before she left Africa on the death of her missionary father she had been told:

> The story is . . . that your mother's father cannot be called a human being . . . there is a race somewhere beyond the forest who are not reckoned to belong to the human family. . . . The natives pretend that they were apes who, in the course of generations, grew into men, but that they lost their way, and now are neither white nor black, man nor beast.[8]

The absence of "soul" particularly distinguished Forest Bokrie from human beings. It is again the absence of soul which is the hallmark of bestiality in the human-like Affri savages of *The Oracle of Baal* (1896).[9] *The Oracle* is a tale in the manner of Sir Rider Haggard of a search for treasure in the recesses of Africa, instituted by Professor Horatio Carmichael, F.R.G.S., and his young friend, Richard Grahame. On the threshold of the land of the Affri they come upon skeletons of savages so shaped that the professor almost had doubts "as to whether the bones had belonged to men or to some very advanced species of anthropoid apes."[10] Following the directions of an ancient will they penetrate farther and farther into the interior. A "Cave of Sleep" through which filters a noxious overpowering gas almost annihilates the travelers. Inside they discover the petrified carcasses of countless animals that have wandered into the death trap, fearful monsters antedating the appearance of man on earth: great lizards, fifty to sixty feet in length, hideous scaled reptiles with serrated spines, and a colossal giant, exceeding twenty feet in height, "the petrified corpse of one of the world's earliest inhabitants."[11]

The Affri turn out to be a tribe of savages, "men in shape only," with faces bearing striking resemblance to one or other of the lower animals. Some have features distorted by nature to resemble apes and baboons, some are like lions and tigers, some like dogs or cattle, but all are to a greater or less degree, human.[12] They possess no articulate language and can reply only by gestures. The Professor cannot classify this race:

> If I had any faith in the Darwinian hypothesis (which I have not) I should be inclined to believe that the Affri tribes belong to a species of very advanced anthropoid apes, brought by the natural course of evolution, to that stage where the brute at length merges into the human.[13]

He conjectures that they may be the long-sought-for missing link, the *Pithecanthropus erectus,* "the animal that stands betwixt the ape and man," and finally decides to rank them with the lower animals: they possess the brute-like faculty of being able to move at will their large flabby ears, and lack that "mysterious essence called by men the *soul.*"

The Affri are slaves to another tribe of more human savages, the Wayangari, who have taught them some of the arts of civilization. In a sudden uprising they destroy their masters and everything that reminds them of civilization, and return to undiluted animalism. The last to go is the faithful Wolf clan. Nine-tenths of the Affri dispose of the only garment their former masters compelled them to wear. A few, hitherto perfectly erect in their gait, begin to crawl on all-fours, "more like apes or baboons than reasonable creatures within measurable distance of attaining to the dignity of humanity."[14] Professor Carmichael, invoking the theory of evolution, cannot explain this reversion to type. To him, evolution "ever climbing after some ideal good," involves successive steps forward up the social scale. The Affri, however, "in defiance of all theories . . . used their newly found liberty merely to take several steps backwards, and slide down to the very bottom of the social scale."[15]

The anthropological romance is merely science well-diluted and well-sugared with imagination. In the work of H. G. Wells the element of science is proportionately larger than in that of the other authors considered. Nevertheless, Wells shares with his age that sense of pride in man's upward climb that made a romanticizing of the Man of the Past inevitable. For Professor Carmichael's conception of evolution as a progressive movement upward, step by step from the slime, and of man as heir of all the ages, was the popular view of the theory which had begun to console the Victorians for the loss of mankind's halo.

"Living fossils" in fiction were more commonly animal than human.* It was difficult to represent a living specimen of a museum man without extravagance or without verging on the ridiculous. The absurd side of such a representation, however, appealed to W. H. Mallock and John Davidson. The search for the "missing link," we have seen, was burlesqued in Mallock's *The New Paul and Virginia*

---

* The surviving prototypes of prehistoric monsters are a lurking danger to the heroes of romance who journey to remote parts of the earth or to other planets. The adventurer in Percy Greg's *Across the Zodiac* (1880) joins the Martians, mounted on riding birds, in a hunt against the pterodactyl or winged reptile. The hero of *A Strange MS Found in a Copper Cylinder* (1894) by James De Mille eludes a prehistoric sea-serpent in the Antarctic, then later, with a race of cave-dwellers, fights scaly monsters like the fabled dragons of old. In A. Conan Doyle's *The Terror of Blue John Gap* (1910) a monstrous animal like a bear yet bigger than an elephant ravages the countryside. It is explained as a survival, in a subterranean cave, of a long extinct prehistoric type of mammoth. Doyle's *The Lost World* (1912) describes the discovery in the South American jungle of the "living dead," iguanodons and carnivorous dinosaurs, dating back in some cases to Jurassic times.

(1878) where it turned out to be a tame monkey, and in Davidson's *Earl Lavender* (1895) where the "link" was discovered to be a gorilla sitting on a prostrate Scotsman. The "primitive" as a character, in his natural setting or lifted out of prehistoric times for burlesque or theatrical effect, took his place in the novel as evolution's most original contribution.

CHAPTER TEN

The Romance of Eccentric Evolution

In romances as old as those of ancient Greece and Rome or of the Middle Ages, the hinterland has always yielded up strange creatures that never were on sea or land. Undiscovered islands in the Pacific, the polar regions, the impenetrable jungles of Africa and South America, have given us Brobdingnagians, Calibans, Anthropophagi and men whose heads do grow beneath their shoulders. What these strange races looked like was dependent entirely upon the nightmarish imagination of the author. But with the establishment of the theory of evolution a definite change became manifest in the nature of these monstrous creations. No longer were they made up as mere fantasies to be just taken for granted. Rather, to give the stories that scientific validity which readers had learned to demand, an attempt was made to represent the bizarre creatures as the result of adaptation to a strange environment, evolution diverging from a straight line, or development resulting from queer mutations. Writers ceased to conjure up mythical centaurs or to resurrect the lost tribes of Israel and the inhabitants of Atlantis.

Under Verne's[1] impetus the romantic search for strange lands on, above, or below the surface of the earth went on with all the enthusiasm of the Renaissance age of exploration and discovery. In strange climes, the romancers knew, evolution would produce strange creatures, off the beaten track of development. Consequently, the evolutionary romance whose setting is some alien land away to the antipodes or beyond the horizon, represents the discovery of freakish instances of aberrant evolution, of quasi-human and semi-human beings.

One of the earliest of these English romances to evolve a bizarre species in an eccentric environment was *Colymbia* (1873).[2] Written by Samuel Butler's medical adviser and intimate friend, Robert Ellis Dudgeon,[3] *Colymbia* describes an energetic subterranean race which has branched off from the human. Somewhat confused in purpose, the work is not pure romance in the manner of Verne. The Colymbians are at times caricatures of humans, at times Utopian idealizations. There is good reason to believe that Dudgeon was indebted, for certain of his satirical passages at least, to Samuel Butler's *Erewhon*, published the year before.

De Courcy Smith, the hero, is shipwrecked near an archipelago of islands of volcanic character in the South Seas. The only survivor, he manages to reach one of the islands inhabited by a race of people who have adapted themselves to a submarine existence, though still capable of existing on land. By a series of ingenious engineering works they are enabled to provide themselves with a complete supply of air in the depths of the ocean. Natural selection has indeed provided them in the course of centuries with some faculties highly useful for life in the new medium.[4] Clothes are considered a superfluity; unrestrained by them, the Colymbians have developed a physical type of singular perfection. The slightest effort enables them to rise and fall like fish, their weight being adjusted to the specific gravity of the sea by certain belts which may be inflated or made heavy.

In a remote age, the Colymbians had taken to the water because of the volcanic nature of the islands and the terrific heat. Their ready faculty of adapting themselves to a subaqueous life led Colymbian philosophers to trace the evolution of the human race from some aquatic ancestor:

> . . . and although all the missing links have not been discovered, it is considered highly probable that an animal allied to the seal tribe was our not very remote progenitor . . . and still farther . . . back, the mollusc,[5]

the instructor tells Smith. Whether their aquatic existence is an atavism or a new development he cannot say. Later (in the chapter entitled "Evolution and Progressive Development") Smith continues the discussion of origins—with a young Colymbian, Julian by name, "an enthusiast for the evolution of species" who thoroughly believed in the continuous advance towards perfection of the human race. Julian regards the Colymbians as the most advanced of human beings and as immeasurably superior to the terrestrial inhabitants. He denies vehemently the descent of man from "those hideous monsters, the apes," and ranges himself with those who maintain our descent from the aquatic mammalia. The seal is his particular choice. In support of his stand, Julian cites anatomical resemblances between man and the seal, and points out that the structure of the human body is better adapted for swimming than for terrestrial life.

De Courcy attacks the Colymbians in their fancied superiority to all other human beings. The Colymbians pride themselves, he suggests, on the paucity of their garments and attribute their physical well-being to the fact; but black-skinned savages wear even less. To this Julian has a ready answer. The black and the white are poles apart as species; for the black savages are derived from the monkeys "to whom they bear a striking resemblance in feature, form and intellect," and were intended for terrestrial life, whereas the white races were constituted for an aquatic existence.[6]

The pragmatic morality of the utilitarian school—the greatest good for the greatest number—is practiced by the Colymbians. They are stanch believers in eugenics, and by law sanction infanticide. Every baby is carefully examined at birth by a committee of learned anatomists and physiologists. If these find it affected with any malformation or deformity or disease which renders it likely that the child will not be capable of making its way in life, it is not reared. The resident director of the Child-Rearing Establishment, points to the singular beauty and healthfulness of this measure:

We merely carry out in a scientific and merciful manner the process performed in a clumsy and cruel manner by nature, which your learned men term 'the survival of the fittest.'"[7]

For the representation of divergent species it was not always necessary to go below the earth as in Verne, or below the sea as in Dudgeon. The opening up of South America, Africa, and the Antarctic by nineteenth century exploration and discovery stimulated speculation along evolutionary lines in these remote sectors. Strikingly varied are the fantastic creations conjured up by the different authors working on the same generalized area.

A remarkable instance of eccentric evolution is illustrated in William Westall's *A Queer Race* (1887).[8] An adventurous search for a lost treasure ship brings the hero, Sidney Earle, to an inaccessible island somewhere off the coast of South America. Miscegenation of blacks and copper-colored aborigines and subsequent mingling of races produced a race with a monkey-like agility, remarkable powers of sight and hearing, and a skin not evenly colored but piebald, splotched red and white or red and black. Every one of these spotted Calibans, as they are called, is a squat Hercules. Their peculiar hue and appearance, Earle speculates, is the "outcome of some subtle evolutionary process, or a mere caprice of atavism."[9]

Not a queer race but a queer individual is the evolutionary quirk described in the novel *Marahuna* (1896)[10] by H. B. Marriott Watson. *Marahuna* centers about the character of a beautiful woman off the beaten track of evolution, whose soullessness brings disaster in its wake. On a scientific expedition to the Antarctic, H. M. S. *Hereward* strangely comes upon a volcanic sector in the polar region where the heat is so intense that flames spring from the waves. Barriers of flame prevent further progress. Suddenly as the ship is about to sail away, the biologist-hero on board, Percy Grayhurst, catches sight of a skiff skimming over the burning water. In it with perfect equanimity stands a stately, golden-haired young woman. She is picked up and taken to London under the guardianship of Percy. She gives her name as Marahuna, but can tell nothing of herself except that she is "not made of humanity's stuff." She is educated like a gentlewoman, but there is a *je ne sais quoi* even apart from her beauty which all wonder at. Finally, Percy realizes that she has no soul; that death means nothing to her, and that she has none of the human feelings of jealousy, pity, fear, hate, or love. Finally, however, there is an upward surge of emotion in her and Percy realizes that it is love, a passionate love for him that has welled up in her.

The biologist's attempts to explain the phenomenon Marahuna on evolutionary principles are carried through the entire novel. Since life evoked under special and peculiar conditions will develop according to those conditions,* Marahuna, the product of evolution in an extraordinary environment, could not fail to be peculiar:

> Man has been evolved from the monad in virtue of the 'promise and potency' therein latent, favoured by a certain definite environment. . . . If then we imagine a monad possessed of an entirely different 'promise and potency' or thrown into a totally different environment, we can quite conceive that the result of the evolution would be very different . . . we might even obtain organised life capable of resisting and enduring flame.[11]

The fact that Marahuna is seemingly human suggests that the general conditions of life in the country of her origin must be similar to those obtaining in the rest of

---

* Following this line of reasoning H. G. Wells describes in his short story *In the Abyss* (1897) the discovery at the bottom of the Atlantic, five miles below the surface, of a race of vertebrated man-like creatures:

> Scientific men . . . see no reason why intelligent water-breathing vertebrated creatures inured to a low temperature and enormous pressure, and of such heavy structure that neither alive nor dead would they float, might not live upon the bottom of the deep sea, and quite unsuspected by us, descendants like ourselves of the great theriomorpha of the New Red Sandstone Age.

H. G. Wells, "In the Abyss," in *The Plattner Story and Others* (London, 1920), p. 91.

the world. That Marahuna is absolutely devoid of human emotions, Grayhurst attributes to the absence in her native habitat of any struggle for existence. But these explanations, the beautiful young phenomenon herself considers specious.

> I am above and beyond human life. . . . Some strange creation, or evolution, you would say, clothing your ignorance in a garb of words.[12]

Under the radically new conditions operating on her in England, Marahuna seems to have become somewhat humanized. Awed at this change Grayhurst begins to wonder at the possibility that "a new evolution in character and nature was being inaugurated."[13] For in the person of Marahuna there is being carried out what his scientific soul once expressed hopefully: "If I had my way I would allow experimentation with human beings. Indeed I would. How else do you think the perfect human being is to be evolved."[14]

The Antarctic is again the scene of the creation of a new species in James De Mille's *Strange MS Found in a Copper Cylinder* (1894).[15] In De Mille's romance a blind race of unique emotional and moral character has come into being, but sharing its environment with prehistoric monsters.

The narrative opens on board a yacht becalmed off Madeira. A copper cylinder is fished up and found to contain a papyrus manuscript. The yachtsmen beguile the tedium of the calm by deciphering what proves to be a tale of the wildest invention.

The hero is one Adam More who, with one companion, gets lost in the Antarctic regions. Attacked by black savages, More loses his companion but escapes in a boat. A swift current sweeps him under a mountain, and along a subterranean stream. In the rapidly widening cavern underground More encounters a great "prehistoric" sea-serpent, but the torrent carries him past before he can be molested. He falls asleep. When he awakes the boat has passed into a new world of an extraordinary character. The light of the strange region is a perfect blaze of moteless splendor, and both the fauna and flora are of the Carboniferous era. The land is peopled by cave dwellers with eyes so feeble they blink in the sunlight.

Morally, these Kosekin, as they call themselves, represent a complete reversal of human standards: their ruling passion is self-denial and the good of others; paupers are the most honored and envied class, and death, the highest blessing. As Adam More had ample occasion to observe, not fearing death the Kosekin are of blind and desperate courage, and seek death the more willingly if they can die for others. Their great joy in life is in heaping gifts upon one another. As fast as wealth is created it is distributed, and consequently the masses are the wealthy class. The poverty-stricken comprise an aristocratic few. When the Kosekin fall in love they separate immediately—not on the principle that absence makes the heart grow fonder, but because the ideal of the race is to "work out the beast." That which distinguishes men from animals is fostered in the Kosekin civilization:

> Animals fear death; animals love to accumulate such things as they prize; animals when they love, go in pairs, and remain with one another. But man, with his intellect, would not be man if he loved life and desired riches and sought for requited love.[16]

Astride giant birds as tall as giraffes, resembling the extinct New Zealand *moa,* these Kosekin conduct a sacred hunt against scaly monsters like the fabled dragons of old. More finally escapes from the land on the back of a flying lizard, an "athaleb."

On board the yacht where the papyrus manuscript is being read, an attempt is made to establish More's tale on a scientific basis. Dr. Congreve, a physician with sound evolutionary knowledge, and Oxenden, an ethnologist, draw heavily on their funds of knowledge to put down the sceptic Otto Melick. The doubting Otto accuses More of having peopled his world with creatures that resemble fossil animals. But the learned doctor insists that there are fossil specimens of animals that still have living representatives. "Think of the recently discovered ornithorhynchus of Australia."[17] Dr. Congreve conjectures that the sea-serpent which More encountered in the subterranean cavern was a plesiosaurus, and that the monsters against which the Kosekin battled on their sacred hunt were primeval saurians: the ichthyosaurus, or fish-lizard, the cheirotherium, the teleosaurus and hylaeosaurus; and More managed to make his escape on the back of a pterodactyl.

Between the ethnologist Oxenden and the physician-evolutionist a debate ensues on the question of the origin of the Kosekin race. The doctor holds the Kosekin to be an original and autochthonous race; but Oxenden, invoking the theory of evolution, argues that they are descendants of the ancient cave-dwelling Troglodytes. Their peculiar eyes he ascribes to their dwelling in caves for generations, just as the fish of the great caves of Kentucky have become eyeless from living in the dark. To their Troglodyte origin, too, the ethnologist attributes the all-pervading love of death held by the Kosekin. The Troglodytes originally regarded death with merriment and pleasure; by the process of evolution this hedonism had developed into its present state, the highest.

Near the end of the strange manuscript the yachtsman who has been reading it aloud, stops, yawns, and tosses the papyrus aside.

Imaginative novelists, exploiting the theory of man's descent from the beasts, found in it inspiration for the creation of hypothetical species and mutations. They proceeded to work forward and backward along the line of evolution, carrying man down toward bestiality, or bringing the beasts up toward humanity. "Reversion ever dragging Evolution in the mud"—the persistent animal in us that keeps dragging us downward and backward—is the theme of novels by Lucas Malet and Marie Corelli; the elevation of the beast toward manhood, the theme of a novel by H. G. Wells.

Lucas Malet (Mary St. Leger Harrison), daughter of Charles Kingsley, uses a specious science in her novel dealing with divergent evolution, *Colonel Enderby's Wife* (1885).[18] The writer prefaces her story with the explanation:

> Even in the most physically and mentally conservative of races there occurs a sudden deflection from the accustomed type. It is probably only a case of reversion, of a return to an older strain of blood. . . . It is a case of this kind which forms the basis of this unpretentious chronicle. Scientifically considered, this is the history of a deviation.[19]

Jessie Enderby, wife of the middle-aged Colonel Enderby, is a glittering young creature of wonderful beauty. There is however something in her character—her light-heartedness, her indifference to questions of right and wrong, her impulsive and fearless nature "untrammeled by conscience, unburdened by the suspense and anxiety which come of spiritual aspiration," which leads Doctor Symes to remark,

> In watching that young lady just now, I could not avoid thinking of the ancient conception of a race of beings supplying the missing link between ourselves and the dumb animals about us.[20]

To him she seems to belong to an "old, old world . . . where nature reigned, and in which man was but the fairest and cleverest of the beasts that perish." Such a Puck-line creature was patently a dangerous mate for a man close to his fifties, and so it proved. Disabled by an accident, his life hardly worth an hour's purchase, Colonel Enderby makes a pathetic attempt to meet his wife's demands for entertainment. The result is collapse, and death. Indifferent to the tragedy, the child of nature, whose soul seems "never to have developed out of the embryonic condition," does not remain a widow long.

Marie Corelli's practice was to take whatever of science she could use, as a springboard for an imaginative flight. Despite her bitter antagonism toward the theory of man's descent, she was not averse in one novel, *Wormwood*[21] (1892), to using the theory to give a pseudo-scientific basis for one of her airy saltations.

The express purpose of *Wormwood* is to show the influence of absinthe on man. Darwin, Miss Corelli points out, traced man's "ascent" from the monkey,

"but had he studied the Parisian *absintheurs* he would most assuredly have added a volume of prophecies for the future to his famous pedigree of the past. For then he would have discovered that absinthe can bring about man's *descent* to the monkey."[22]

The "Boy," or "bete" as he is sometimes called, son of André Gessonex, an artist and absintheur, is Miss Corelli's model of what absinthe can do. The youth's race strain was fouled from the start, the author intimates, for his grandfather was an evolutionary scientist, an atheist who "talked of Matter and of Force,—of Evolution and of Atoms."[23] After the savant's suicide his son André, the boy's father, took to absinthe. The boy himself is a perfect example of the Stone Age, with the countenance of a mingled monkey and savage.

He would do credit to the antediluvian age, when man was still in the process of formation. The chin . . . is not developed,—the forehead recedes like that of the baboon ancestor,—the nose has not yet received its intellectual prominence. . . . He is an animal, made merely . . . 'to arise, kill, and eat.'[24]

This snarling creature, André calls "mon singe."

H. G. Wells's scientific romances were written in the afterglow of Thomas Henry Huxley's inspired teaching at the Normal School of Science.[25] The mood in which he wrote *The Island of Dr. Moreau* (1896)[26] was strongly impregnated by the Grand Elucidator's influence.

The evolutionary philosophy which underlies the story of the beast-men was contained in an article Wells wrote the same year, "Human Evolution, an Artificial Process."[27] This article developed tangentially certain ideas put forward in Huxley's famous Romanes lecture, "Evolution and Ethics." Wells developed the thesis that man intrinsically is as much a savage today as was the Palaeolithic brute. The difference between civilized man and his brute ancestor is an artificial factor inculcated in man after birth by the social example and precept of the civilized body into which he was born a member. Morality in this view is simply the padding of social and emotional habits by which society keeps the round Palaeolithic savage in the square hole of the civilized state. A discord between man's innate palaeolithic disposition and the artificial factor imposed on it was suggested as a new phrasing for the moral conflict. This evolutionary view of sin H. G. Wells tried to convey in *The Island of Dr. Moreau.*[28]

The Island of Dr. Moreau is based on the assumption that the gulf between animality and humanity is so small that it may be bridged with a surgeon's knife. The problem that Dr. Moreau attempted to solve was: Can we, by surgery, so accelerate the evolutionary process as to make man out of a beast in a few days or weeks?

In the person of Dr. Moreau, Wells presents a science-crazed physician, inspired to a series of hideous experiments by the revelations of evolution. By grafting the tissue and bone of animals in a long series of operations over many years, Dr. Moreau succeeds in evolving a race of half-humanized brutes to whom he teaches speech and a peculiar moral law to stifle their predatory tendencies.

The "Law" of the Beast-Men, Wells would suggest, is the social veneer that covers the essential bestiality of Moreau's creations, and humanizes them. So long as the "Law" has vogue no creature is harmed. Prendrick, the narrator of the story, who came to the island by accident, wanders into the den of the Beast-Men and is mistaken by them for one of their number. He is told that he must learn the "Law." The Ape-Man enforces his command with a threat. Prendrick joins in the insane ceremony led by the "Sayer" of the Law, repeating the idiotic formula line by line:

> Not to go on all-Fours; *that* is the Law. Are we not Men?
>
> Not to suck up Drink; *that* is the Law. Are we not Men?
>
> Not to eat Flesh or Fish; *that* is the Law. Are we not Men?
>
> Not to claw Bark of Trees; *that* is the Law. Are we not Men?
>
> Not to chase other Men; *that* is the Law. Are we not Men?
>
> And so from the prohibition of these acts of folly, on to the prohibition of what I thought then were the maddest, most impossible and most indecent things one could well imagine.[29]

Moreau, Prendrick realizes, has even infected their dwarfed brains with a kind of deification of himself. But the "Law" is powerless to interpose any sufficient breakwater against the wild surge of passion, or to provide any adequate protection against the tiger leaps of besetting sin. The "Law" is broken and Dr. Moreau falls prey to one of his creatures. With Moreau dead, and the pressure of the taboo reshaping their psychology removed, the Beast-Men little by little begin to drift back to their congenital natures like the "missing links" or Affri of *The Oracle of Baal*. Their articulate speech becomes lumps and sound, they shed their clothes, drink by suction, and begin to crawl on all-fours. The "stubborn beast-flesh" grows back day by day soon to obliterate all traces of humanity.

Civilization—a system of taboos—serves to countervail the downgrade tendencies of animal temperament. But man is at base a beast, Wells intimates. He underlines this thesis in representing the narrator Prendrick's return to England in a passage strongly reminiscent of the final pages of *Gulliver's Travels.* Once more on native soil after his experiences among the Beast-People, Prendrick becomes subject to a strange delusion. He thinks of the people he meets not as men but as beast-men who may some day forget the "Law," and reverting, drop on all-fours and bare their teeth. "I go in fear," he says,

> I see faces keen and bright, others dull or dangerous, others unsteady, insincere; none that have the calm authority of a reasonable soul. I feel as though the animal was surging up through them; that presently the degradation of the Islanders will be played over again on a larger scale. . . . I would go out into the streets to fight with my delusion, and prowling women would mew after me, furtive craving men glance jealously at me, weary pale workers go coughing by me with tired eyes and eager paces like wounded deer dripping blood. . . . Then I would turn aside into some chapel, and even there . . . it seemed that the preacher gibbered Big Thinks even as the Ape Man had done.[30]

The civilization of men who trace their pedigree from an ape-like ancestor is bound to differ from that of men who trace theirs from archangels ruined. The fatalistic shadow of *damnosa hereditas* broods over man in *The Island of Dr. Moreau* underlining the closing words of Darwin in his *Descent of Man:*

> With all his noble qualities . . . with all these exalted powers—man still bears in his bodily frame the indelible stamp of his lowly origin.[31]

The quirks of evolution under special and strange conditions, these romances show, could be used to justify as marvellous human phenomena as any fine frenzied imagination might body forth. Whether for the creation of Apollos or Calibans, the evolutionary principle was equally facile. By natural selection or the application of scientific selection in the romance, a super race or a degenerate one might be developed. That life evoked in a remote and peculiar environment will evolve according to that environment, is a basic principle in the creation of the bizarre creatures of these romances. The exotic beings of the Antarctic and the seabottom are described in accord with this formula. Where the principle is not adhered to, the author yielded to the temptation to represent the eccentric race in satirical contrast to our own. In subsequent development, the satirical romance as a genre is represented most frequently in association with the evolutionary romance of the future, culminating in the Juvenalian work of Aldous Huxley, *Brave New World.*

CHAPTER ELEVEN

Evolution and the Idea of Progress

The ideas most often set forth in evolutionary romances constituted a debate concerning human destinies. Evolution itself, it must be remembered, does not necessarily mean, applied to society, the movement of man to a desirable goal. It is a neutral scientific conception compatible either with optimism or with pessimism. According to different estimates it may appear to be a guarantee of steady amelioration or a cruel sentence. It has been interpreted in both ways. On the one hand the optimists declared that the implications of evolution held forth the promise of a golden age in which a race of supermen would develop. Many thoughtful and thoughtless people were ready to discern—as Huxley suggested—in man's "long progress through the past, a reasonable ground of faith in his attainment of a nobler future." On the other hand, the pessimists declared that man has ceased to develop, that the species is not permanent, and that the earth if not otherwise destroyed will some day fall into the sun. They shared the not very sanguine views of Huxley in his late years: ". . . even the best of modern civilization appears to me to exhibit a condition of mankind which neither embodies any worthy ideal nor even possesses the merit of stability."[1]

Darwin himself gave impetus to the optimistic view of evolution in the conclusion to his *The Origin of Species.* There he had remarked:

> As all the living forms of life are the lineal descendants of those which lived long before the Cambrian epoch, we may feel certain that the ordinary succession by generation has never once been broken, and that no cataclysm has desolated the whole world. Hence we may look with some confidence to a secure future of great length. And as natural selection worked solely by and for the good of each being, all corporeal and mental endowments will tend to progress toward perfection.[2]

But the ablest and most influential development of the argument from evolution to Progress was the work of Herbert Spencer.[3] With a confidence not less assured than that of a Condorcet or a Godwin, Spencer argued that human nature is subject to the general law of change, and that the process by which human nature slowly but continuously tends to adapt itself more and more to the conditions of social life, points to an ultimate harmony. This principle of social evolution and gradual psychical improvement of the race is, in Spencer's Synthetic Philosophy, a necessary sequel to the general cosmic movement toward order and completeness.

In the Victorian age, an age of prosperity and self-complacency, it was inevitable that an optimistic interpretation of evolution should be readily received. Neither Spencer nor Darwin was, however, directly responsible for the spreading of these glad tidings. Since the general reading public touches nothing more serious than the popular novel, or magazine, or the daily newspaper, the novel had to serve the function. A class of popular novelists was ready at hand, eager to perform the work of keeping the lower levels posted with the news of what was stirring higher up.[4] These writers undertook with enthusiasm the task of subtly flushing the imagination of men with the consciousness that they were living in an era which, in itself vastly superior to any of the past, need be burdened by no fear of decline or catastrophe, but, trusting in the boundless resources of science, might securely defy fate.

The idea of evolutionary progress is set forth in a series of philosophic and romantic novels. The philosophic novels, assuming the continuing advance of modern man, represent him as evolving along moral, aesthetic, and mental lines. The work of Charles Kingsley, George Meredith, and Samuel Butler will be seen representing respectively these three fields of progress. The romantic novels, somewhat Utopian in character, depict the progress of evolution in strange worlds under strange conditions—with current and futuristic implication; or the progress of evolution in the future when men and machines will have been developed to perfection. All these have, in common, faith in the perfectibility of mankind.

Charles Kingsley's devotion to religion and science, which led him to seek and find a facile means of reconciling the two, has already been touched on.[5] His *Water Babies* (1863), in addition to suggesting the groundwork for such reconciliation, contains extended illustration of Kingsley's moral philosophy based on the principle of progressive evolution. Somewhat fantastically the theory is presented that the transmutation of species from the beginning of time was dependent on the possession or lack of moral qualities.

Strictly within the tradition of Lewis Carroll and Stephen Leacock, Charles Kingsley produced *Water Babies* during the term he served as professor of modern history at Cambridge. He wrote it nominally for his youngest son Grenville. Of the various opinions entertained about the fantasy, one seems to be unanimous; viz., that the book cannot be generally understood by children.

Water Babies is an *olla podrida* of Darwinian implications. The purpose of the tale seems to have been to adapt Darwin's theory of the natural selection of species to the understanding of children, by giving it a moral and religious as well as a scientific application.

The fairy tale is in two parts. Part One tells with much pleasantry how little Tom, the chimney sweep, employed by the bully Mr. Grimes, runs away, falls into a river, and is turned into a water-baby. Part Two describes Tom's "move upwards, working out the beast"—his journey as a water-baby from the river to the sea to the Other-end-of-Nowhere, during which he makes intimate acquaintance with all sorts of aquatic creatures and learns from them the principles of moral evolution.

With his tongue in his cheek and one eye cocked mischievously at the empiric scientists, Kingsley pursues the implications of Darwinism to the creation of the new species, "Hydrotecnon Ptthmllnsprtsianum, or some other long name like that,"—the water-baby. Sceptics like Cousin Cramchild and Aunt Agitate might shake their heads; savants of evolutionary science like Professor Owen, or Professor Sedgwick, or Professor Huxley, or Mr. Darwin might raise their eyebrows; but Kingsley insists with the sobriety of the scientific logician that land babies may reasonably change into water-babies.

> For if the changes of the lower animals are so wonderful, and so difficult to discover, why should not there be changes in the higher animals far more wonderful, and far more difficult to discover.[6]

Tom has the misfortune to be caught in the net of the great naturalist, Ptthmlhsprts, professor of Necrobioneopalæonthydrocthonanthropopithekology; but the eminent professor (who "got up once at the British Association and declared that apes had hippopotamus majors in their brains just as men have") is completely nonplussed by the new species and affirms solemnly that he does not exist. But to the satisfaction of the reader and generations of youngsters unborn, Kingsley has proved that he does exist. The professor, driven insane by a glimpse of this species unaccounted for by his studies, proceeds to compose treatises proving the moon to be made of green cheese. As for Tom, the water-baby, having learned from the beasts which perish what is morally right and what morally wrong, he is pronounced victor "in the great battle, and fit . . . to be a man."[7]

In the great watery world into which the chimney-sweep-become-water-baby swims, Kingsley found ample scope and illustration for his pseudo-scientific elements. He chose the watery world principally because he knew it so well, and because the number of transformations which go on in it are so large, and so easily capable of a semi-moral significance. It served best to illustrate his purpose. For example, the specific difference between salmon and trout Kingsley interpreted as a difference between enterprise and industry on the one hand, and stupid greediness on the other, leading to evolution along divergent lines—as shown in this conversation between his water-baby and the salmon:

> "Why do you dislike the trout so?" asked Tom
> "My dear, we do not even mention them, if we can help it; for I am sorry to say they are relations of ours who do us no credit. A great many years ago they were just like us; but they were so lazy, and cowardly, and greedy, that instead of going down to the sea every year to see the world and grow strong and fat, they chose to stay and poke about in the little streams and eat worms and grubs; and they are very properly punished for it; for they have grown ugly and brown and spotted and small; and are actually so degraded in their tastes, that they will eat our children."[8]

Again, the Gairfowls are meant to be the types of races who die out through mere traditional pride, from refusing to avail themselves of fresh blood, and determining to stand all alone on the precedents and etiquettes of ancestral usage. The last of the Gairfowls preens herself on her conservatism in not sprouting wings:

> They must all have wings, forsooth, every new upstart sort of bird, and fly. What can they want with flying and raising themselves above their proper station in life? In the days of my ancestors no birds ever thought of having wings, and did very well without; and now they all laugh at me because I keep to the good old fashion.[9]

The same moral Darwinism is the idea of the story of the idle Doasyoulikes whose history over a span of thousands of years fairy Ellie shows Tom. It is evolution in reverse which the water-baby beholds, a devolution of man to the level of apes as a result of his bestial behavior and failure to do what he did not like. "Folks say now that I can make beasts into men, by circumstance, and selection, and competition," the fairy tells Tom; ". . . and, if I can turn beasts into men, I can, by the same laws . . . turn men into beasts."[10]

The same general drift is intended to pervade all of the "Tom-fooleries." Wherever moral qualities or the germs of moral qualities begin, there, at least, is a turning point of natural development or degradation in the individual, and thence also in the species. Indeed, all the various physiological transformations in the story are designed to illustrate some such notion as this.

The Autobiography of Christopher Kirkland (1885) by Mrs. Lynn Linton, has already been discussed in connection with the effect of the theory of evolution on the hero's faith.[11] It is significant likewise for the evolutionary philosophy which Christopher Kirkland acquires; an evolutionary philosophy which was Mrs. Linton's own, for the novel is in large measure her autobiography curiously inverted by her assumption of a masculine character.

Kirkland concerns himself chiefly with the moral aspects of evolution. Despite the exceptions which religious leaders were insisting on, he holds firmly to the thesis that the moral sense is as much a matter of evolution as are the physical and intellectual elements in man. It is his faith that further evolution will proceed principally along moral lines: "It has taken millions on millions of years to evolve man out of protoplasm; it will take some thousands more for all the savage and the beast to be educated out of him."[12] The Law of Progress becomes his religion; and he fervently prophesies that society will evolve a world where reason, self-control, and altruism will replace the passions, appetites, and selfishness as motive forces.

Marie Corelli had made it a practice in her novels to attack Darwinism and modern science as synonymous with animalism and atheism. But she finally came round to an acceptance of evolution in so far as it fitted into her religious conceptions. Unlike Mrs. Linton who looked forward to the evolution of man to perfect humanity in an earthly state, Miss Corelli envisaged the evolution of man to the status of an angel in paradise.

Our long evolution from "prehistoric slime and loathliness," she tells us in *The Master Christian* (1900),[13] has come about because there lies within us God though in the germ. Man's earliest instincts taught him to climb toward God. If this were not true, man would be still where he began, "in the woods and caves,—an uncouth savage with nothing save an animal instinct to lead him."[14] But we must be constantly reminded of our glorious destiny, for man too often tends to creep back again on all-fours to the days when he was a beast merely.

> We are always forgetting that we have been, and still are in a state of Evolution,—out of the Beast God has made Man,—but now He expects us, with all the wisdom, learning and experience He has given us, to evolve for ourselves from Man the Angel—the supreme height of His divine intention.[15]

Again in *Temporal Power* (1902)[16] Marie Corelli expressed her faith in the direction which God has given to man's evolution. In response to Paul Zouche's expostulation that there is no purpose in the world, that the universe is merely a "chance whirl of gas and atoms," the heroine Lotys, "the soul of an ideal," exclaims:

> There is an Intelligence at the core of Creation! . . . study the progress of humanity. Do you not see that while the brute creation remains stationary, (some specimens of it even becoming extinct), man goes step by step to higher results?[17]

Marie Corelli's faith in the upward spiritual progress of the species was, on a different plane, shared by George Meredith. But Meredith looked forward hopefully to the advance of body and brain as well, in this future race of aristocrats. In his work and that of Grant Allen, there is set down the path man will have to tread to achieve this mental and aesthetic progress.

Though George Meredith has been variously referred to as the "poet of evolution,"[18] in his novels there is no evidence that he had more than a most general knowledge of the principles of evolution. In all his writings there is no mention made of the name Darwin.[19] Passing references to the theory of evolution and to Darwinism are to be found in his novels, but none that would suggest an assimilation of their ideas. An explanation of such indifference may be suggested by the passage in the Prelude to *The Egoist* (1879), in which Meredith rejects Science and accepts the Comic Spirit as the specific for our modern malady of sameness:

> We drove in a body to Science the other day for an antidote; . . . and Science introduced us to our o'er-hoary ancestry—them in the Oriental posture; whereupon we set up a primaeval chattering to rival the Amazon forest nigh nightfall, cured, we fancied. And before daybreak our disease was hanging on us again, with the extension of a tail. We had it fore and aft. We were the same and animals into the bargain. That is all we got from Science.

Nevertheless, for all his insistence that "We have little to learn of apes," Meredith did admit a surprisingly large slice of evolutionism into his transcendental philosophy of body, brain, and spirit. He was a strenuous believer in the progress of the race and the upward march of humanity. Meredith saw humanity as the flowering of nature and the mind of man as the flowering of humanity. In the development of brain and intelligence, he believed, the process of evolution was still at work. This, of course bespoke a buoyant and sun-lit optimism. Through the lips of Diana of the Crossways he asked significantly, "Who can really think and not think hopefully?"

Meredith's hope for the future progress of man was tied up with his trinity of body, brain, and spirit. The acme of human evolution results from an aesthetic, mental, and spiritual development. In the representation of Sir Willoughby Patterne's courtship of eighteen year old Clara Middleton, in *The Egoist,* Meredith gives expression to his creed of aesthetic evolution—Darwin's theory of sexual selection.

Sir Willoughby congratulates himself that Clara, who has money and health, beauty and breeding, has chosen him as mate above the many others who have been competing for her hand. "A deeper student of Science than his rivals," the Egoist boasts to himself that Clara's choice is the working of the principle of natural selection.

> We now scientifically know that in this department of the universal struggle, success is awarded to the bettermost. You spread a handsomer tail than your fellows, you dress a finer top-knot, you pipe a newer note, have a longer stride; she reviews you in competition and selects you.[21]

Meredith's hope for the future of the race resides in the doctrine thus facetiously presented. He underlines the intuitive character of the selection of the fittest:

> She cannot help herself; it is her nature, and her nature is the guarantee for the noblest race of men to come to her. In complimenting you, she is a promise of superior offspring. . . .
>
> Thus did Miss Middleton acquiesce in the principle of selection. . . .
>
> He looked the fittest; he justified the dictum of Science. The survival of the Patternes was assured.[22]

There need be no doubt that George Meredith completely espoused the Darwinian principle of sexual selection, as he expressed it in *The Egoist*. It formed an integral part of his dream of "the creation of certain nobler races now very dimly imagined." There is comic irony of course in Sir Willoughby Patterne's representing himself as one of the fittest, and seeing in his choice of and by Clara Middleton after he has "aired his amiable superlatives" in her eye, the working of the law of natural selection. But Meredith invoked his Comic Spirit to ridicule not the natural law but the Egoist who had the presumption to apply it to himself.

Grant Allen was a skilful member of the group of scientific middlemen who applied themselves to the popularization of the doctrine of evolution. He had studied philosophy and physical science, and formed an evolutionary system of his own, based mainly on the work of Herbert Spencer. Though he made occasional forays into the field of original research, his chief function as he saw it was to give plain people an opportunity of forming for themselves a judgment worth something on the subject before them, instead of feeling themselves forced to bow to the *ipse dixit* of a man who knows how to use a microscope or a scalpel. This he did in a series of popular scientific articles, always with an evolutionary moral—his *Vignettes from Nature* (1881), *The Evolutionist at Large* (1881), *Colin Clout's Calendar* (1883), etc. Frequent were the tributes paid by Darwin, Spencer, and Wallace to the value and accuracy of Allen's work, and to the original contributions he made to theories which he popularized. On reading *Vignettes from Nature,* Huxley was moved to write Allen:

> If Falstaff had been soaked in Evolution instead of sack, I think he might have 'babbled o' green fields' in some such genial fashion. I am not quite sure whether you will take this as a compliment or not; but it is meant for a great one.[23]

Late in life Grant Allen took to fiction. His novels usually rest on or exemplify some principle in science like heredity, racial instincts, or some phase of evolutionary science. Of little or no literary value, these novels are interesting in their reflection of Victorian tastes in fiction, and in their facility to "write down" scientific thought to appeal to these tastes.* In *A Splendid Sin* (1896)[24] Allen develops, as had Meredith, a scientific philosophy of love on the basis of Darwin's thesis of sexual selection.

Hubert Egremont, a virile, healthy youth, is a physiologist by profession. Hubert is a firm believer in the predominant significance of heredity, established by his studies at hospital and asylum; likewise he has a particular interest in the theoretical aspects of evolution. It is on this belief and interest that the novel rather melodramatically turns. Despite his many inquiries, Hubert has been kept in ignorance of his father. His mother merely intimates that the father was a man of noble attainments. At a hotel bar Hubert has a chance meeting with a military gentleman, a Colonel, whose drinking habits arouse his professional interest. The Colonel frankly admits his weakness as he gulps down glass after glass of brandy; and confides to Hubert that his grandfather died at eighty, his father at seventy, and that he expects to go at sixty. "It's an interesting example of what Darwin calls the law of accelerated inheritance," he announces. Hubert analyzes his case: alcoholic narcosis on the way to insanity; then tells the concierge with dramatic irony: "His grand-

---

* Allen's *The Typewriter Girl* is an example of this light, "scientific" fiction, with chapters entitled "The Struggle for Life," and "Environment Wins." The heroine, Juliet Appleton, unemployed, finally gets a position as typist and stenographer. She draws the appropriate evolutionary moral:

In the struggle for life I had obtained a footing. . . . I had proved myself fittest by the mere act of survival. . . . The sole remaining question was, Could I adapt myself to my environment? If so, I had fulfilled the whole gospel of Darwinism.

Grant Allen, *The Typewriter Girl* (New York, 1900), p. 22.

children will be born idiots or epileptics . . . if that man has a son, the son is doomed to insanity before thirty." Shortly after there is a revelatory scene in which Hubert discovers that the drunken Colonel is the husband of his mother—his father. Immediately his prophecy of the inheritance heredity has given the son of such a father strikes his brain. But Grant Allen was no Ibsen. The tragedy of *Ghosts* is averted with the confession by Hubert's mother that, though married to the dipsomaniac, long years before she had met her affinity—a great American poet—and committed the splendid sin.

Hubert's disquisition on Darwinian sexual selection gave courage to his mother to reveal her *grande passion*. In conversation with his uncle, Sir Emilius, Hubert had developed a doctrine of Goethean affinities founded on an extension of Darwin's theory. This, his mother had overheard. Love, Hubert argued, is the voice of Nature telling us clearly what mate the powers that rule the universe have fashioned for us.

"It is only the most beautiful, the most perfect, the most agile, the most effective that get selected on the average, and such selection increases in the long run the beauty, agility and effectiveness of the species."

"Still, all this reasoning amounts to no more than what Darwin taught us long ago," Sir Emilius insisted.[25]

But Hubert emphasized the originality of his contribution to the theory of natural selection. Just in proportion as we rise in the scale of being, he affirmed, we find at every fresh grade a further narrowing down of the possible range of choice and of attraction. Among the lower animals any mate will suffice; with the higher, aesthetic preferences begin to come into play, and give us at last such visible results as the plumage of the peacock, the bird of Paradise. So, too, among men. With the savage almost any squaw is as good as another; the rustic begins to demand, at least, physical beauty; higher cultivated types are progressively fastidious.

On such grounds, Hubert denounced marriage that admits any other consideration save that of immediate fitness. And on the developing eclecticism in marriage which he detected in the evolution of species, he rested his hope for the aesthetic superiority of the coming race.

Like Grant Allen, Samuel Butler received accolades for his extensions of the Darwinian theory; but unlike Grant Allen, Butler received his posthumously. During his lifetime, the author of *Life and Habit, Evolution Old and New, Unconscious Memory,* and *Luck or Cunning?* was treated with contumely by professional biologists.[27] Lacking portfolio as a scientist, Butler, whose interest in biology was primarily philosophic, found the scientists' doors closed against his brilliant speculations in the field of evolution. *Life and Habit* (1877) was received as a gigantic joke, and even Grant Allen joined the chorus of frogs in croaking at Butler's *Evolution Old and New* (1879),* as "leaving the reader without a single idea upon any subject whatever."[27] But Butler had his day. In 1909, seven years after his death, in a memorial volume published on the occasion of Darwin's centenary, Butler was to receive from an official biologist of the first rank full recognition. Professor William Bateson therein lauded Butler as "the most brilliant and by far the most interesting of Darwin's opponents, whose works are at length emerging from oblivion."[28]

Before Butler began in 1873 the composition of his third and last novel, *The Way of All Flesh,*[29] he had ini-

---

* There is irony in this. In *A Splendid Sin,* Hubert Egremont is made to remark, in discussing the omnipotence of heredity: "I almost feel at times as if I had no individuality at all of my own. I recognize myself as nothing more in the end than the sum of my joint parental tendencies." (Grant Allen, *op. cit.,* p. 52.) This is palpably Samuel Butler's favorite doctrine of the physical continuity of parent and offspring set forth in *Life and Habit* and rehearsed in *Evolution Old and New.*

tiated his career in speculative evolution. For twelve years he worked intermittently at the novel, not intending to publish until all who could be hurt by it had died. *The Way of All Flesh* was published in 1903 after Butler's death, but the author had not touched it after 1885.[30] During the period of the novel's gestation Butler put out three of his four works in evolutionary controversy, *Life and Habit, Evolution Old and New,* and *Unconscious Memory.*

The theory contained in *Life and Habit* is basic to a complete understanding of *The Way of All Flesh.* The novel is admittedly an illustration of theories which Butler had settled upon after renewed study of evolution and its implications, and which he incorporated into *Life and Habit.* With the passing years Butler had become more and more convinced that evolution must be divorced from Darwinism. Darwin and his followers upheld evolution and denied design. Butler's view in *Life and Habit* posited both evolution and design. Butler's concept of design in evolution held the cause of variation in plant and animal life to be effort, purpose, struggle of the organisms themselves for higher adaptation to conditions or bodily modification, made necessary by change in environment or circumstances. Design is intrinsic in the organism's will to live.

The inheritance of acquired habits is, of necessity, an integral part of Butler's theory. The principle of "unconscious memory" he advanced to account for the transmission of these acquired habits: the principle adumbrated in *Erewhon.* By the unconscious memory of an act or series of acts that has been repeated incessantly we have learned all that we know; and through us life has learned all that it knows from the amoeba up to man, or earlier still, from the primordial cell.

The specific propositions in *Life and Habit* which Butler was to treat imaginatively in *The Way of All Flesh* are four: First, that there is a *bona fide* oneness of personality existing between parents and offspring up to the time that the offspring leaves the parent's body. Secondly, that in virtue of this oneness of personality the offspring will remember what has happened to the parents so long as the two were united in one person, subject, of course, to the limitations common to all memory. Thirdly, that the memory so obtained will, like all other memory, lie dormant until the return of the associated ideas. Fourthly, that the structures and instincts which are due to the possession of this memory will learn, in the course of time, to act and develop without self-consciousness. Such a phenomenon of heredity as reversion to a remote ancestor, exemplified in *The Way of All Flesh,* and countless other troublesome questions are answered by these premises.

The Way of All Flesh tells the story not so much of a character as of a family organism, insinuating the nature of the organism in the name Pontifex.[31] Butler visualized the Pontifex family with its strong hereditary characteristics transmitting the Pontifex cell almost without modification for four generations. The saga follows the family slowly and successively emerging out of business into the Church, stagnating there, then emerging out of the Church into the free air of creativeness and self-determinism.

The life history of the hero, Ernest Pontifex, begins with his great-grandfather, John Pontifex, a carpenter. Self-sufficient, happy, this industrious old man is Ernest's first and only agreeable progenitor. The generations swing on, past George Pontifex the domineering publisher, past Theobald his son, who is bullied into taking orders and jockeyed into marriage with the smug, self-satisfied Christina; finally to Ernest, their child, who, in complete subjection to his parents, suffers so cruelly from the pharisaical tyranny of his father during childhood and school days, that after his ordination, reaction from suppression begins. The catastrophe is sudden. He makes improper advances to a young woman whom he takes for a prostitute, and is sentenced to six months' imprisonment. It is a profitable time of reflection and mental growth. On emerging, ruined, from prison he casts off Theobald and Christina for a fresh start in life. He marries Ellen, a former maidservant in the family—a disastrous union—and sets up a second-hand clothing shop. His fortune is now up, now down, but he is on his own. Finally with the help of a bequest from his Aunt Alethea he comes through his various experiences a happy and dignified human being.

In *The Way of All Flesh* the narrator remembers Ernest's great grandfather, the village carpenter; he has known about the rise of the grandfather, George Pontifex, to wealth, and has been at school with the father, Theobald. He is thus able to see the family in perspective and to estimate the importance of inheritance.

> If a man is to enter the kingdom of Heaven (he remarks), he must do so, not only as a little child, but as a little embryo, or rather as a little zoosperm—and not only this, but as one that has come of zoosperms which have entered into the kingdom of Heaven before him for many, many generations.[32]

This view would seem to determine the career of Ernest Pontifex before his birth, but that is not the effect of his life as presented to the reader. Butler as an evolutionist, we have seen, was no longer a follower of Darwin, but of the French biologist, Lamarck. With Lamarck, Butler held to the theory (presented in *Life and Habit*) that species arise not from the accumulation of minute accidental differences, but from the exercise in nature of a function corresponding to the human will, arising under a sense of need. Ernest Pontifex is a weak character in an unfavorable environment—but the events which seem to spell his ruin shock him into an exercise of his latent will.

Ernest's bringing up and education by no means fitted him for survival. Theobald and Christina had reared their child not with an eye toward those abilities which prepare for life, but to evolve those "virtues convenient to parents." And these last sprang from nothing real, from no empirical doctrine, but were mystically conceived from the inexorable and artificial moral code of father Theobald and the hallucinations of mother Christina. Ernest's boyhood, schooling, and ordination proceeded under this superimposed morality, which, like the system in *Richard Feverel,* held up under the strain until nature burst disastrously through it to get at the fresh air of reality. Ernest's misdemeanor, trial, and imprisonment, and consequent loss of profession, stripped him of those incrustations which had insulated him from life. Now bared to the stinging blasts, a sense of need assailed him. He had broken with his past; he must make his own way, independent of that past. That his long-drawn out effort to find his own soul and become captain of it was successful, was a matter however, compounded not only of this modern Hamlet's will, but also of his ancestry.

In his memoir on Samuel Butler, Henry F. Jones informs us that though Butler was aware of the length of the first few chapters, the necessity in his mind of emphasizing Ernest's prenatal experiences precluded his curtailing them.[33] Since the climax of Ernest's life is bound up in a reversion to the primitive strength of his great grandfather, such stress was not amiss. This ancestor, John Pontifex, was a healthy, versatile, and thriving growth with a store of strength upon which he had never drawn. Through two generations it had remained untapped by his descendants until Ernest found himself thrown on his own resources. The gentility which the Pontifexes had taken so much pains to acquire and had bequeathed to Ernest, he threw away in this crisis as an excrescence, and returned to his original nature for the vigor to adapt himself anew.

Of course he read Mr. Darwin's books as fast as they came out and adopted evolution as an article of faith. "It seems to me," he said once, "that I am like one of those caterpillars which, if they have been interrupted in making their hammock, must begin again from the beginning. So long as I went back a long way down in the social scale I got on all right."[34]

The crisis in Ernest's life which put him down for a time was a spur which started him fighting upward on an altogether new tack. This divergence was Ernest's contribution to the hereditary stock. Had he continued in the old ways, following a profession which he detested, obeying the dictates of parents whom he despised, Ernest would have carried on his effete family line to inevitable extinction. Father Theobald had stifled his own original tendencies, submitted to his father (Ernest's grandfather), and in his own person prolonged his father's nature unchanged. Ernest's nature, at first, was merely an extension of his father's and grandfather's. Whatever energy the virile John Pontifex had contributed to the stock, disuse had caused to stagnate. In Butler's phrase, the family stream of memories, inadequate to deal with new conditions, was about to run dry. But the exigency of struggle stirred up Ernest's will to live and with it those dormant memories of strength, born of his great-grandfather's own struggles and successes. The result of this break with the immediate past was a new departure, the creation of a new set of memories which Ernest transmitted to his own children.[35]

Just such an evolutionary moral Butler reads into the supermuscan efforts of a fly to reach the lip of a cup containing hot coffee. Unwittingly the fly had alighted on the thin skin of milk on the surface of the coffee. Perceiving its danger it made heroic efforts to reach the edges of the cup, and finally did. The narrator comments: "As I watched him I fancied that so supreme a moment of difficulty and danger might leave him with an increase of moral and physical power which might even descend in some measure to his offspring."[36] Just such an accident had set Ernest struggling supremely to gain his own feet and shaken "him out of himself by shaking something else into him."[37] But it was by his own will that he snapped the strings which were holding him to the path that led to degeneration. This "revolt of the life force" was in Butler's philosophy what made for the upward climb of species.

Towneley, a rowing blue and an athlete, genial to the "nobodies" like Ernest, cultivated by everyone, yet without a trace of snobbishness, not fastidious in his pleasures (as witness his visits to "the governess" of shady morality, Miss Snow), yet unsparing in his efforts to save Ernest from humiliation, is for Butler the *beau ideal* of the organism which has succeeded in adapting itself to the circumstances of life. Towneley has common sense and he has right instinct, and by virtue of these inestimable possessions, albeit without either knowledge or the desire for it, he can be relied upon to take the right path with as much success and confidence as Ernest, with twice his mental equipment, can be trusted to take the wrong every time. Towneley is therefore Ernest's foil.

> The people like Towneley (Ernest sadly reflects), are the only ones who know anything that is worth knowing, and like that, of course, I can never be. But to make Towneleys there must be hewers of wood and drawers of water—men, in fact, through whom conscious knowledge must pass before it can reach those who can apply it gracefully and instinctively as the Towneleys can.[38]

Here we have Butler's biological doctrine, and his practical application of it in a nutshell. Human beings, like all other organisms, are at different levels of evolution. Those who are most advanced are not those who have the most brains, but those who have already learned in the persons of their ancestors that art of living which their less fortunate brethren are toilsomely acquiring in their own; and who, once having learned it, do not know that they have it, but possess it unconsciously. People of this type get right unfailingly and instinctively what others learn laboriously and consciously with many mistakes and much expenditure of intellectual effort. Brains then are only a half-way house to the acquisition of right instincts. Ultimately, when we do everything right instinctively we shall have no need to think at all. Then we shall be happy. Intellect is thus for Butler an evolutionary make-shift. It does laboriously and clumsily what instinct does quickly and infallibly, and each advance in evolution witnesses a fresh suppression of intellect by instinct.[39] In this matter Butler was of the school of Hardy, and anticipated the Freudian emphasis of the unconscious.

The Way of All Flesh thus may be considered the graphic illustration of those ideas which are basic in all Butler's works. The origination of evolution in a sense of need and exercise of will to fill that need, the inheritance of memory and stimulation of that memory by associated ideas, the idea of intellect as an evolutionary makeshift and of unconscious memory as the consummation of intellect, were repeated, embellished, and expanded from work to work. In his return to the Lamarckian hypothesis assuming the existence and function of will in creation, we mark Butler's chief point of departure from Darwinism. The distinction between Darwin's theory and Butler's is of importance because it corresponds to a difference between the English realists and the French naturalists. The latter reflect the determinism implicit in Darwinism; to them the individual is the resultant of heredity and environment—predestination is a scientific fact. English realism, on the contrary, in general reflects the freedom of the will which, illusion or not, is basic in our sense of experience.[40]

Butler's work in this direction was that of an optimist, a believer in progress. The author put faith in nature, bent on its own improvement under a sense of need. He tried to dissipate the deadening atmosphere emanating from the materialistic consequences of the Darwinian theory and its suggestion of haphazard, blind, unintelligent forces. In *The Way of All Flesh,* the moral is to promote the ideal of an upward moving humanity, embodied in Ernest Pontifex's spiritual odyssey, his long drawn out effort to seek his proper place in the scheme of things.

Not so sure a faith in progress as that held by Butler, but as firmly founded in knowledge and understanding of Darwinian doctrine, was the faith held by Mrs. Olive Schreiner, author of *From Man to Man.*[41] Without blinking the ruthless destruction and monstrous waste along the path of evolution, Mrs. Schreiner was able to extract from the theory some measure of assurance of the upward sweep of humanity.

Mrs. Schreiner's *From Man to Man* was published posthumously in 1926, six years after her death, forty-three years after her more famous *A Story of an African Farm.* Actually, Mrs. Schreiner ceased work on *From Man to Man* as far back as 1910 or 1911, having begun writing it as early as 1873, after reading Darwin's *Descent of Man.*[42]

From Man to Man details the tragic lives of two sisters, Rebekah and Bertie. Without attempting any fusion Mrs. Schreiner has interpolated through the story speculations on Darwinism which appear in random jottings by Rebekah, a sincere student of botany and geology. Whenever Rebekah can find time from her manifold household duties, she turns to her study. There amid her fossils and botanical specimens, debating with herself, she spins out her lengthy thoughts on the survival of the fittest which are duly recorded in her notebooks.

Rebekah was no pessimist. She nursed the hope that some day all that was good in man would be embodied in a future god-like race. How that could be brought about became the subject of her musings. At first thought she contemplated the most drastic solution: suppressing, cutting off, and destroying the less developed individuals and races, leaving only the highly developed to survive. But the inevitable questions occurred: whence would come the people to judge others? Wouldn't racial, social, and other forms of bias influence these judges? And even if there were impartial umpires, how would the standards of those desirable and undesirable to the race be drawn up? What are the prime requisites that would make a person fit for this future race? Physical or mental? Or what combination of the two?

The conception that the weeding out of the unfit might best be left to nature herself provoked Rebekah to fierce indignation. By nature's way the strong prey on the weak and the fittest survive. The fittest?—certainly not in the sense in which humans use the word, Rebekah insists.

> The fittest has survived! Under water, half buried in mud, only the outline of the jaw and two deep slit eyes show where the alligator lies. Age after age he has lain in the mud and slime.
>
> The boa contrictor wakes in the morning, and before night bird and beast have been crushed in its mighty fold; it lies stupefied and torpid with the creatures it has consumed in its expansive inside. It has survived them, not because it is fairer or higher in the scale of being than they but because so greasily and silently it could creep upon them. The cobra strikes dead man and beast, and survives, not because he is braver or higher or even stronger, but because beneath that tooth she carries that little poison bag and strikes so silently and it may be in the dark.[43]

Nature's plan has made for the survival of those most fit to destroy. It has killed off the winged reptile and a thousand noble complex and brilliant forms of life, and has saved the crocodile and the python. Age after age it has killed off among advancing humans, creatures who to impersonal ends sacrificed the arts of destruction and self-defence. The only strength which it directly preserved is predatory strength.

In the terrible conflict for survival the victory has gone to the strongest jaw, the longest claw, and the biggest belly. But, Rebekah questions, in the strife has not the beautiful been preserved—singing birds, flowers, the wonderful intelligence of man and beast? Can it not be argued that the conquest by the most cunning, the most merciless, the most consuming, the muscularly or osseously stronger may have added to the keenness of the eagle's eye, the length of the springbuck's graceful bound to flee from the jaws of her enemies? It may have sharpened the wits of all creatures who have had to escape.

> It may even have added to that intensity of anguished love which makes one baby mierkat try to drag a smaller away to safety when it sees the hawk approaching, because the little people have learnt by a long racial experience what the claw and beak mean, and those who have loved and aided each other most have survived—the fittest to live, not the fittest to kill in that case![44]

Here Rebekah was ready to take her departure from Darwinism. Love was a law of life and survival of which natural selection took no account. Yet everywhere mother-love and the tender nurturing of the weak underlies life, and the higher the creature the larger the part it plays. The insect who fastens her eggs to the top of the highest bough where the tender shoots will first sprout; the bird who draws the soft down from her breast to warm the nest, the mammal feeding her young from her breast—"through all nature life and growth and evolution are possible only because of mother-love."[45] Man individually and as a race is possible on earth only because, not for weeks or months but for years, love and the guardianship of the weak over the strong has existed. Love to Rebekah was the exception in the rule of nature which for perfecting life demands the destruction of the weak by the strong.

It is this law of love, sympathy toward the weak, which Rebekah carries into her own life. When her little son Frank refuses to walk with the colored girl Sartje because his friends scoff at him, Rebekah draws from evolution the appropriate moral. Vividly she unfolds for Frank and her other children the long pageant of life on earth from the first appearance of shell-fish up through the coming of man, dwelling on the weakening of selfishness, the strengthening of sympathy. One race if higher, Rebekah contends, must help the other up. No race must be regarded as "inferior." Take from us, the superior race, all we have gained and learned from other races and peoples, and we should go back and back until at last we come to the

> . . . little cave mother with her baby tied by a skin on to her back, peeping out at the door of the cave to see if the man with his bone hook or flint arrows was coming home with game.[46]

Through the lips of Rebekah, Mrs. Schreiner developed the ideas set forth at the time the book was first begun, by Herbert Spencer and Henry Drummond. It was Herbert Spencer who first laid down a law of altruism—struggle for racial as well as for individual life; and Henry Drummond's *The Ascent of Man* (1873) rechristened it the law of Love.

CHAPTER TWELVE

Evolution and the Idea of Degeneration

Evolution, we have suggested, lends itself to a pessimistic as well as to an optimistic interpretation. While some have tended by natural disposition to dwell with pride on the elements of harmony and progress, others have been led to attend more to the element of chance and conflict.

In the early years of the post-Darwinian era, the gloomy interpretation of the Darwinian doctrine was adopted by clerical apologists who were desirous of condemning the new theory by showing up its repellent dreariness. An identical view was taken by opponents of theology who, assuming the truth of the theory, sought by means of it to demonstrate the untenability of the old idea of a perfectly benevolent Creator. The processes by which Darwin explained the survival of the fittest were far from affording any indication of the design of some kindly transcendent breeder; or from implying imminent effort in the animal. Since all variations were attributed to chance, and the sorting out to the mechanical action of the environment, mind was banished from the universe. A fatalistic philosophy was the inevitable by-product of such doctrines.

Darwinism, however, even aside from its teleological implications, was soon seen to nourish the seeds of pessimism. We stand in the midst of a wonderful and cruel world, a world moreover in which pain and cruelty, the slaughter of the weak by the strong, or their decay and death by their own imperfect organization, are not accidental defects, but are of the very law and essence of the development of life on the globe. So far as life and the improvement of life are the outcome of the struggle for existence, the organic world seems to have its roots in suffering. This was the aspect of the struggle for existence which Huxley underlined in his Romanes lecture of 1893. Animate nature, he held, was like a vast gladiatorial show, a Hobbesian warfare of each against all, the result of which is merely the survival of the most suitable, not of the best in any sense. He pointed to egoism, rapacity, and a bloody despotism of the strong over the weak as the qualities of survival value in the transmutation of species. "For his successful progress as far as the savage state man has been largely indebted to those qualities which he shares with the ape and the tiger."[1] Against the cosmic process thus villainously and sensationally presented, Huxley urged war, to perpetuate those values which men deem highest. "The practice of that which is ethically best—what we call goodness or virtue—involves a course of conduct which, in all respects, is opposed to that which leads to success in the cosmic struggle for existence."[2]

Such a generalization was well adapted to feed the stream of pessimistic thought. Weighted with all the authority which in these days attaches to scientific deliverances, and stamped with the cachet of great name, evolutionary pessimism gained added prestige. Further, it was remembered that the Darwinian notion of the struggle for existence had been derived not from the cosmos but from the more virulent form of human competition. It was man's war against man as Malthus had represented it that recalled to Darwin's mind the war of plant against plant, of beast against beast. Man, it seemed, though he had traveled the farthest along the path of evolution, had not traveled far enough from his animal ancestry. He was still stirred by the same fierce impulse to survive at the expense of his fellows, still moved by strength to tyrannize over the weak. The same rugged individualism of the ape and tiger were manifest in the Victorian world of fierce competition, pitiless appetite for wealth, and idolatry of material things. Indeed, the ruthlessness of the beast was seen to be at times more intense in man than in the lower orders. The desires of an ape and a tiger are limited; if they can get sufficient food for the day they are content for the day, and the struggle is only for the day. The desires of a rational being, however, are practically unlimited, and every object may become an object of desire. The competition between him and his fellows is practically unlimited, and the struggle for existence has no ending.

Those who survived under such conditions might be termed the fittest, but they hardly would be termed the best. The struggle for existence, it followed logically, may have as its outcome degeneration as well as amelioration: evolution may be regressive as well as progressive. In this view the future of mankind was far from a roseate one. "The theory of evolution encourages no millennial anticipations," Huxley had said.[3] And Tennyson in *Maud* had given voice to the same feeling of despair:

> As nine months go to the shaping of an
> infant ripe for its birth
> So many a million of age have gone to the
> making of man:
> He now is the first, but is he the last? Is
> he not too base?

The literature of regressive evolution or degeneration represents man as essentially an animal, endowed with consciousness and intellect that serve simply to make him aware of the cruelty of the world and the futility of existence; the product of chance that, having called him to life to strut his tiny hour upon his tiny stage, will usher him out—perhaps to be succeeded by another species. Man's life on earth is represented as largely a battle in which the beautiful and humane and fine natures are driven to the wall by the small-minded, the mercenary, the rapacious, or more ruthlessly efficient.

The years when Thomas Hardy began writing were a decisive turning-point in the mental life of the nineteenth century. The great intellectual event of the age dividing its chronology as with a water-shed was the publication of Darwin's *Origin of Species.* Hardy's earlier poems were written some half-dozen years after the appearance of the *Origin of Species;* and his first novel was published in the same year as the *Descent of Man* (1871). Darwin's proof of the evolution hypothesis and the progress of the exact sciences gave authority to the naturalistic philosophy which was rapidly becoming an accepted mode of thought. As a young man Hardy had been among the first to acclaim the *Origin of Species;*[4] and the general movement of his thought thereafter followed the rhythm set by his age. The idea of design was consigned to the limbo of exploded myths, and the principle of natural selection and its twin brother, the principle of chance variation, enthroned in its place. The result was Hardy's conversion to a deterministic monism.[5]

Thomas Hardy's reading of evolution gave no hope of man's being able to control his destiny. Too large a part in life is played by chance. Hardy saw consciousness as an adventitious circumstance in the cosmic process, something for which nature had made no provision. In 1883 he wrote in his diary: "We human beings have reached a degree of intelligence which Nature never contemplated in framing her laws,* and for which she consequently has provided no adequate satisfaction."[6] The same fortuitous element which had brought man's intellect into existence Hardy saw at work plaguing man's life. Flukes and accidents are constantly occurring to thwart man, rendered vulnerable to disappointment by his insatiable appetite for happiness.

---

* George Gissing opposes Hardy in this, refusing to believe that chance controls and that consciousness evolved purposelessly:

The flower simply a flower, and there an end on't? The man simply a product of evolutionary law, his senses and his intellect merely availing him to take account of the natural mechanisms of which he forms a part? I find it very hard to believe . . .

George Gissing, *The Private Papers of Henry Ryecroft* (New York, 1918), p. 159.

Life's tragedies, to Hardy's mind, come of man's anthropomorphic reading of nature. Nature is by no means so concerned with man as an individual as he is himself. She is, indeed, indifferent to man's desires and aspirations. But man, expecting nature to mold herself to his will, finds himself balked. His personal tragedy, born of his own misconceptions, he lays at nature's door.

Hardy viewed man from both ends of the telescope. From the one, man loomed large, dominating the scene; from the other, he shrank into insignificance against a cosmic canvas. The tragedies of man's life, large as they may seem to man himself, and fraught with deepest consequences, are when viewed *sub specie aeternitatis* but a mote in a sunbeam. It is this ironic contrast of the greatness and littleness of life which gave the tang of bitterness to all of Hardy's novels.[7]

The obvious instance is *Two on a Tower* where the characters are shown against the whole stellar universe, for ironic commentary on the pettiness of their human drama. Just as in that novel Hardy revealed the insignificance of man in space, so in *A Pair of Blue Eyes* (1873)[8] he pointed the insignificance of man in time, suggesting thereby the ephemeral nature of the human species.

In one of those many accidents to which Hardy gave so much weight Henry Knight, hero of *A Pair of Blue Eyes,* geologist and man of letters, loses his footing at the edge of the "Cliff, Without a Name," and hangs for some minutes on the verge precariously clinging by a tuft of sea pink. Lost to all but the curving face of the cliff round and beneath him, and prompted by an imbedded fossil whose eyes "dead and turned to stone" stared at him from the rock, Knight relives the whole cycle of organic evolution on the earth. The fossil was one of the early crustaceans called Trilobites, separated from him by millions of years; but Knight's mind swept back momentarily to the age which gave it birth, then forward and back again slowly through the varied scenes that had had their day between this creature's epic and his own.

Time closed like a fan before him. He saw himself at one extremity of the years, face to face with the beginning and all the intermediate centuries simultaneously. Fierce men, clothed in the hides of beasts, and carrying for defense and attack, huge clubs and pointed spears, rose from the rock like phantoms before the doomed Macbeth. . . . Behind them stood an earlier band. No man was there. Huge elephantine forms, the mastodon, the hippopotamus, the tapir—antelopes of monstrous size, the megatherium, and the myledon—all, for the moment, in juxtaposition. Further back, and overlapped by these, were perched huge-billed birds and swinish creatures as large as horses. Still more shadowy were the sinister crocodilean outlines—alligators and other uncouth shapes, culminating in the colossal lizard, the iguanodon. Folded behind were dragon forms and clouds of flying reptiles: still underneath were fishy beings of lower development; and so on till the lifetime scenes of the period confronting him were a present and modern condition of things.[9]

Prophecies of the passing of man from the earth, implicit in the kaleidoscope visualized by Hardy's geologist, are again the obsession of the hero in Percy White's *Andria* (1896).[10] So bedeviled by the notion of the passing of man from the earth is Louis Otway, hero of White's *Andria,* that he flings away a reputation to drive his prophecy home. Famous as the author of *Science and Civilization,* Otway suddenly suffers a change of philosophy and becomes a misanthrope. He speaks veiledly of a revelation soon, "a great secret—a secret compared to which Darwin's discovery fades into insignificance."[11] Man, he announces is "an ephemeral insect on a slowly dying planet" whose end will be "as contemptible as the beginning when, a hairy apelike monster, he first walked erect in the gloom of the forest."[12] In article after article he preaches his thesis, and anticipates the coming of a more endowed breed. Finally, after threatening the public with a book to prove that man, whether "under the hairy hide of the anthropomorphous arboreal ape of Darwinism" or the dress clothes of civilization, was merely a "morbid and useless excrescence on the face of nature," he commits suicide.

The thinness of the line that marks off man from the beast, and the possibility of bringing the beast over the line into humanity is the basis of a story of regressive evolution, *The Curse of Intellect* (1895),[13] by Frank C. Constable. Constable's novel lugubriously suggests that the beast would learn to resent his humanization and long for the carefree joy of the forest.

The story turns on the mesmeric ability of Reuben Power, a gifted intellectual but a profound cynic. Power stigmatizes all happiness that does not proceed from the mind. Reason, he holds, places man above all other beings; and reason therefore should be developed, even at the expense of happiness. Following his credo, he renounces the society of man whom he loathes to pursue his studies.

When next he appears he is accompanied by a monkey whose education he has undertaken on the evolutionary principle that "man's boasted intellect is but cut from, and carved . . . from physical nature."[14] Power has taught the monkey to live as an ordinary human being, and to converse with him. The sounds the monkey makes are a painfully grotesque parody of men's speech, and convey intelligible meaning only to Power. But otherwise the Beast, as he is called, is altogether civilized; even so far as not to permit talking to go on when there is any music and to glare indignantly at a joke about Providence or God. On occasion he acts as host at Power's receptions, invitations to which have been signed "Semnopithecous Rubicundis."

It is finally learned that Reuben Power trained the monkey in the art of civilization in order to find out what a beast like that might think of mankind. Himself a deep-dyed misanthrope, Power retained a sufficient scientific detachment to want a criticism of man from an unprejudiced point of view. But he stacked the cards, and in time the beast began to look at man with Power's eyes. Soon there is apparent an inhuman expression of hopeless misery in Power's face. The Beast has become master.

The Beast takes up the pen to carry out at least Power's wish that he register a monkey's view of the world. He remembers vaguely his aimless but heedless life in the jungle before Power found him and subjected him to the whip; and the painful struggle that went on in him between instinct and intellect. Then he awoke one morning able to think and will. But the road was not always upward. Power was teaching two monkeys, but with a huge rock the Beast mashed the skull of the other. However, intellect finally triumphed over instinct. In slow succession there followed articulate speech, reading, writing, and the worship of man's achievements, as recorded in books. Then the actual world was opened to the Beast but with the devil at his elbow in the shape of Power's cynicism. Man's meanness, his worship of wealth and position, his selfish pride soon engendered in the Beast a contempt for man equal to Power's.

Nothing to me shows more strongly the curse of intellect than the fact that in every country vast herds of men are trained to killing as a profession, and that these men are not only honoured, but form the really best class.[15]

He loses all sense of gratitude toward Power who has torn him from nature and "reduced him to the level of man." Their positions change; the Beast is now the whip-holder. Finally, driven to despair by the dread that Power, in calling forth intellect, has perhaps too drawn a soul into him to suffer in a future world, the Beast strangles his erstwhile master, then puts a bullet through his own head.

Power's friend, Colin Clout, apostrophizes on the end: "How . . . painful for the poor beast, with the power of reflection suddenly born in him, full, from reading, of belief in man's godlike greatness, to be confronted suddenly with the human beast as he is.[16]

There is little but the deep indigo of his mood to mark Constable's post-Darwinian Beast from Peacock's pre-Darwinian Sir Oran Haut-ton. Essentially, however, Sir Oran was a creature of instinct, a noble savage, whereas the Beast was a creature of intellect, however speciously evolved.

The process of evolution which has produced man has not only failed to grind out of him the element of bestiality, but in its cruel and indifferent course seems destined to grind out of him those elements which men hold highest; worse, to preserve in him those elements which men deem lowest. The triumph of the fittest does not necessarily mean the triumph of those that should survive. It means rather the survival of those that fit in best with prevailing conditions. Since nature is amoral, the struggle for life can terminate in the victory of types unfit to live. "Some may survive by carrying a disease which will destroy other species more energetic or more fit to live."[17] By a rugged individualism, brutality, or ruthlessness, others may perpetuate their type, leaving in their wake the altruistic, the idealistic, and the beautiful.

Lucas Malet wanders from her main theme in *Colonel Enderby's Wife* (1885)[18] to anathematize those who would find justification in Darwinism for their pitilessness in human relations.

Ah! what an inspiring and consolatory doctrine is that of the survival of the fittest. How agreeably it strengthens the hands of the capable, merciless, strong, and causes the gentle and timid weak to duck under. How beautifully it is calculated to increase the exercise of the more robust virtues—pride arrogance, cruelty. . . . And . . . foolish persons . . . will still cherish a hope of the survival of an unfit minority, among whom it may remain possible to cultivate gentleness, modesty, and a quiet love of personal liberty, without being immediately trampled underfoot.[19]

George Gissing's *New Grub Street* (1891)[20] is concerned not with man's inhumanity to man which provoked Lucas Malet's outburst, but with the inhumanity of the inexorable law of survival. The theme is italicised in a remark made by the successful Jasper Milvain: "The struggle for existence among books is nowadays as severe as among men."[21]

New Grub Street paints the struggle both among books and among men in the literary world of Gissing's day. Natural selection plays the role of villain; for the adaptation of writers to their environment and to prevailing conditions is the basis of the action. The general conception is that in the modern Grub Street success is assured only by adopting the most frankly utilitarian and mercenary views. The author must write what the public wants and is willing to pay for, without consulting his artistic conscience; and he must employ every art of self-advertisement and of acquiring influential friends in order to win out in the literary game. Those authors who can change their writings to suit the fickle public taste for the new will survive and live, secure in the thought that tomorrow will bring their breakfast, luncheon, tea and dinner. Others, those poor souls with ideals who have set themselves rigid rules of composition, can exist only so long as the public favors their particular brand of literature.

Illustrating Gissing's theme are the two most prominent characters, Edwin Reardon and Jasper Milvain. Edwin Reardon is a highly temperamental and conscientious artist who refuses to sacrifice his artistic ideals to the public taste. So long as the public relishes the classical turn which Reardon featured in his books, the critics applaud, thus assuring a large sale. But when the public whim shifts, Reardon declines to shift with it. The thought that he can make no progress whatever is before him night and day. Hampered by poverty and by his wife's lack of sympathy with his resolute stand against cheapening his art, he struggles to turn out novels that his wrecked brain can no longer conjure. When one is complete, he receives no more than a pittance for it; for the public will have none of his work. He finally has to submit. He parts from his wife and, rapidly disintegrating, dies, a premature victim of privation and despair.

In Jasper Milvain, foil to Reardon, we have a character who, in the Gissing sense of the word, is fit for life—mercenary, selfish, and unscrupulous. Milvain, a clever reviewer and essayist, frankly accepts the materialistic standards of the New Grub Street, and wins social and literary success. While from every viewpoint neither his work nor his character was the equal of Reardon's, nevertheless, in the eyes of nature he was better fitted to inhabit the earth.

Engaged to Reardon's widow, secure in an editorship, Jasper Milvain hears of the suicide of Harold Biffen, another victim of the struggle for survival.

> There are two of my companions fallen in battle. I ought to think myself a lucky fellow, Marian. What?
>
> You are better fitted to fight your way, Jasper.[22]

The few novels we have considered, though holding in common a somewhat disillusioning view of man's relation to the beasts that perish, view man with varying degrees of jaundice. Thomas Hardy, with the objectivity of a scientist, viewed man as a passing phase in cosmic evolution. Hardy is alone in his scientific detachment. The attitude of the other authors is more touched with emotion. Percy White, with a morbid disdain for man and his fancied superiority, hailed his ultimate departure from the universe with enthusiasm. From wells of pessimism equally deep Frank C. Constable drew his black-hued conception of the worthlessness of human intellect. More realistic by far in its approach was the evaluation of evolution's supposed benefits by Lucas Malet and George Gissing. To them evolution, blindly sifting, too frequently chooses the chaff and discards the wheat.

CHAPTER THIRTEEN

The Coming Race and the Idea of Progress

By attempting to establish the laws and to trace the lines marking the upward climb of species to the present, the evolutionary sciences evolved a fairly rational scheme of the past. From this it was not a far cry to the idea of using scientific knowledge like a two-edged knife to cut forward into the future as well as backward into the past, for an account of the descendants as of the descent of man. The novel followed the stroke of the knife cleaving backward into the past—the romance of the prehistoric past—and forward into the future—the Utopian romance. By far the more popular vista was that into the future; the picturing of the Man of the Future, that mysterious being who will look back across a dim gulf of time upon imperfect humanity of the nineteenth and twentieth centuries with just such kindly and half-incredulous scorn as we now condescend to bestow upon our own club-wielding, ape-like ancestor. This, with its obvious appeal to the imagination, gave rise to what may be called the fantasies of science, developed with great popular success by Jules Verne and H. G. Wells.

Glimpsing man's future in the magic mirror of his past, these evolutionary romances unfold a picture of a perfect race, living in complete harmony with its environment—a real Eden in the future that would more than equal the storied ones of the past.

> Every tiger madness muzzled, every serpent
> passion killed
> Every grim ravine a garden, every blazing
> desert filled.

As prophecies of the future they do not seem altogether preposterous like the myths of Plato and More. Though idealism is there, it is idealism working—along evolutionary lines, based on attainable facts, and mapped out in accordance with the natural order or sequence from existing conditions, and duly regarding the limitations of nature, both human and physical.[1] Some, failing to realize to the full the implications of the evolutionary principle, conceive of a hard and fast state of perfection, no longer subject to the stresses and strains of progress. This is true, chiefly, of the early Utopias, following the example of Bulwer-Lytton's *The Coming Race*. The more modern Utopias are more scientific. They aim to represent not a perfect state of society but a progressive one. The thing to strive for, they suggest, is social progress not social perfection; incessant becoming, not stagnant being.

The Modern Utopia must be not static but kinetic, must shape not as a permanent state but as a hopeful stage, leading to a long ascent of stages.[2]

Utopia in these imaginative romances is situated either in the present or in the future. In the former case it is discoverable in some strange and hitherto undiscovered land on or below the earth, or in some extra-terrestrial world, usually Mars. In all of these romances, however, there is the suggestion of the path that the evolution of the future will follow. In all of them, to give point to the relationship of mankind, the "other world" is peopled by a race human in its origins, if not altogether human in its development.

Man in this ideal world is debrutalized, a being in whom the animal forces are absolutely domesticated and ancillary to the human. The mechanism of natural selection has been supplemented by an ethical process for the elimination of the unfit. This race of aristocrats is in complete command of complex machines, whose invention and discovery have proceeded at an accelerated pace. Science, too, has continued its unprecedented development and with its aid Utopians are able to harness the universe to their will—always for the greatest good.

Bulwer-Lytton's *The Coming Race* (1871)[3] was one of the earliest pieces of fiction to exploit the theory of evolution and predict the development of man into a "calm, intellectual race." In the preface to the second edition of *Erewhon,* the author, Samuel Butler, takes cognizance of the many little points of similarity between his book and Bulwer-Lytton's *The Coming Race* which antedated *Erewhon* by several months.[4] So strong are some of these resemblances that only Butler's assurance that his work was composed entirely independent of the other precludes suspicion of plagiarism. But, since the inspiration for both was derived from the evolutionary views then paramount in the mind of all, the identity in the two novels seems not remarkable but natural.

In *The Coming Race,* the hero, an American, discovers a Utopian people living in the bowels of the earth. Their marvelous civilization is attributed to a fluid, *vril,* very much akin to electricity, which can at will stimulate, strike, cure or kill, and gives illimitable power. The narrator learns by degrees that the forbears of this race descended through caverns into the earth during inundations many thousands of years before the time of Noah, when the land was being turned into the sea and the sea into land. For a time the race of the inner world made the sort of discursive progress that the races of the surface are now making. But the discovery of *vril* brought such confusion to an end. The Vril-ya, or members of the regenerated society, explain their transcendent superiority by the struggle for existence in which only the highest organizations are preserved by natural selection. There is a legend among them that the race was driven from the upper world

. . . in order to perfect our condition and attain to the purest elimination of our species by the severity of the struggle our forefathers underwent; and that . . . we are destined to return to the upper world and supplant all the inferior races now existing therein.[5]

The women have the upper hand in this advanced society. Owing to their greater powers over the mysterious fluid *vril,* they have undoubted physical superiority. Likewise, they are the chief cultivators of speculative science for which they are fitted by the greater delicacy of their intellects.

Chief among the anatomical modifications which distinguish the Vril-ya from the human race is that of the hand; their large, massive thumb (with as great a difference from ours "as there is between the thumb of a man and that of a gorilla") and a visible nerve, not present in the earliest progenitors of the race, account for their superior use of *vril* power. Further, they make use of wings (not a part of their body but detachable) which have become as natural to them as arms or legs. An infant wills to fly as intuitively and unconsciously as he wills to walk. It is this Lamarckian or Butlerian will which has been responsible for that superiority which the Vril-ya have over their super-terrestrial neighbors. Thus, the will to fly, transmitted from parent to child generation after generation, has at last become an instinct. By such means the Vril-ya have accelerated their evolution beyond belief, since their *vril* civilization is a matter of only about 7,000 years.

From a giant frog, the Vril-ya trace their lineal descent; and quote from the sayings of their Philosopher-forefather, "Humble yourselves, my descendants; the father of your race was a *twat* (tadpole)."[6] The anatomical and analogical agreements between the An (man; plural, Ana) and Frog were established by naturalists by the same arguments that Victorian scientists offered to prove the kinship of man and ape. With satire of obvious application, Lytton tells of the heated controversy that went on seven thousand years before in the Wrangling

Period of History over the question of the descent of the An. One school of philosophers asserted the An to be the perfected type of the Frog, another, that the Frog was the highest development of the An. Frogs were palpably superior to the Ana according to this latter school, for Frogs were hairless and

> . . . the object of the higher races of the *Ana* through countless generations has been to erase all vestige of connection with hairy vertebrata, and they have gradually eliminated that debasing capillary excrement by the law of sexual selection.[7]

Among the Vril-ya are no poverty, no crime, no misery, no government except a benevolent patriarchal rule, and no war—for with *vril*, war would mean the annihilation of civilization. The sciences are highly developed, though the arts are held somewhat in contempt—for the somewhat Platonic reason that art gives expression to the violent and disorderly passions which have disappeared among the Vril-ya. There is nothing but the calm pursuit of knowledge by well-regulated minds. Mankind is reduced to a dead level with no heroes and no criminals.

Calm, intellectual, effortless, and somewhat monotonous is the civilization of the Vril-ya. By contrast with the fretful warring states on the earth's surface, their land seems Utopian. And to Lytton who based his ideal society on the thesis that "mortal happiness consists in the extinction of strife and competition," the civilization of the Vril-ya was Utopian. Theirs was "the coming race" to whose serene tranquillity we humans must turn as an ideal; for societies constituted like ours cannot improve but "tend to further deterioration."

The appearance of *The Coming Race* in the same year that saw the publication of Darwin's *The Descent of Man,* was a mere coincidence. It was the material of the *Origin of Species* (1859) and the numerous controversies started by it that stimulated Bulwer-Lytton to speculate about the evolutionary development of man. Successive evolutionary romances derived their inspiration and sustenance as much from *The Coming Race* as from *The Descent of Man.*

The Man of the Future is usually contemporary. We of today come upon him in strange worlds adjacent to our own, or receive him as a visitor from a strange planet. In Grant Allen's *The British Barbarians* (1895),[8] he steps out of the twenty-fifth century into the present. From that distant time when man has become a thoroughly rational being, as calm and intellectual as Lytton's Coming Race, an ethnologist, conversant with the recorded aspects of nineteenth century life but desirous of first-hand knowledge, visits the nineteenth century very much as a modern ethnologist might visit Zululand. Reincarnated in the highly respectable suburb of Brackenhurst, England, the scientist, Bertram Ingledew, regards the Philistine society of Brackenhurst as a highly advanced stage of barbarism held together by conventions and prejudices which he describes as savage taboos. "It's only in Europe, where evolution goes furthest, that taboo has reached that last silly pitch of injustice and absurdity,"[9] he tells Freda Monteith, wife of the town's Crœsus. For everywhere he turns he comes upon customs that remind him of the savage practices of Polynesia or Swaziland: taboos about property, costumes, the relations between the sexes, manual labor, class distinctions—all unreasonable. England dare not boast of civilization with so many vestiges of the uncouth primitive in her institutions.

Ingledew's frank, aristocratic bearing makes him seem like a civilized being in the midst of barbarians. With the sweet reasonableness of the perfectly rational being, he cannot be moved to anger or excitement; but that the man of the twenty-fifth century is as impulsive in the presence of a beautiful woman as the man of the nineteenth century, is amply demonstrated by his relations to Freda

Monteith. He has no superstitions, and accepts no rules of conduct or morals which have not stood the test of reason. His guiding principle is to question all things.

Ingledew's own innate superiority to those around him leads Freda Monteith to venture half-heartedly the conjecture that the ethnologist of the future is a being from another planet. But life on another planet, the scientist insists, would not be human life or animal life as we know it.

> Our race is essentially a development from a particular type of monkey-like animal,—the Andropithecous of the Upper Uganda Eocene. This monkey-like animal itself, again, is the product of special antecedent causes, filling a particular place in a particular tertiary fauna and flora, and impossible even in the fauna and flora of our own earth and our own tropics before the evolution of those succulent fruits and grain-like seeds for feeding on which it is specially adapted. Without edible fruits, in short, there could be no monkey; and without monkeys there could be no man.[10]

Consequently, Ingledew implies, being human he must of necessity have evolved not on another planet but on this one; and in his person he holds out the prospect of what human evolution, following reason and reasonable goodness, may create in the future.

Shot by the irate husband of Freda Monteith for violating the taboo that hedges marriage, Ingledew fades back into the twenty-fifth century, carrying Freda with him.

A new departure in Utopian romance is marked in George Du Maurier's *The Martian* (1897),[11] which exhibits the coalescing of the Gothic and the scientific elements. Taking cognizance of this synthesis of disparate factors, Miss Dorothy Scarborough stated:

> The application of modern science to supernaturalism, or of the supernatural to modern science, is one of the distinctive features of recent literature. Ghostly fiction took

a new and definite turn with the rapid advance in scientific knowledge and investigation in the latter part of the nineteenth century, for the work of Darwin, Spencer, Huxley, and their co-laborers did as much to quicken thought in romance as in other lines.[12]

Three years before the publication of *The Martian,* George Du Maurier set down in *Trilby* an optimistic sentence which was to become the theme of the later novel. Ringing out the old order and ringing in the new, he remarked,

> It was a good time in England . . . a time of evolution, revolution, change and development . . . a keen struggle for existence—a surviving of the fit—a preparation, let us hope, for the ultimate survival of the fittest.[13]

The Martian gives a prospective view of the nature of these "fittest" in the half-light of illusion and through the veil of the supernatural.

Du Maurier's third and last novel tells the story of the handsome, gifted, strangely idealistic Barty Josselin. Passing from one discouragement to another and on the verge of going blind, Barty contemplates suicide. Upon waking from what he intended to be his last sleep, he finds a note in his own handwriting but signed "Martia," discussing his secret thoughts and advising him where to find a physician to cure his blindness. Martia is a Martian who has reincarnated herself in Barty as a sort of *alter ego,* or guardian spirit.

For many years she communicates with him constantly and helps him. As Barty sleeps, Martia dictates to him books reconciling science and faith, and novels which have "as wide and far-reaching an influence on modern thought as the *Origin of Species,* that appeared about the same time."[14] His writings achieve marvelous reforms in morals and eugenics: the race becomes on the average four to six inches taller, and no couple dare marry until they are pronounced physically, morally, and mentally fit.[15]

Martia reveals something of life on Mars. Man in Mars is, it appears, a very different being from what he is here. He is amphibious, and descends, like the Colymbians, "from no monkey, but from a small animal that seems to be something between our seal and our sea-lion."[16] According to Martia the beauty of a Martian is to that of the seal as that of a Greek god is to that of an orang-outang. His fingers and toes are webbed, his senses extraordinarily acute. In addition to our five senses, the Martian possesses a sixth sense, a sensitivity to magnetic currents which enables Martians to read one another's minds. Since this made deception impossible, the moral sense of the Martians has progressed far in advance of ours. They wear no clothes but the exquisite fur with which nature has endowed them, and which enhances their transcendent beauty. All useless and harmful and mutually destructive forms of animal life that mar the ideal have been cleared away.

But this extra-terrestrial Utopia is not in a state of static perfection. By virtue of its long and patiently evolved magnetic sense the race is "still developing towards perfection with constant strain and effort."[17] This, despite the fact that the planet is far advanced in its decadence, and within measurable distance of unfitness for life of any kind.

When a Martian dies, his soul flies toward the sun; but on its way thither it may well stop over to dwell in the soul of some inhabitant of another world. Many Martian souls come to the earth in this way, and incarnate themselves in the most promising newly begotten, unborn earthlings. Martia herself came to the earth before she had lived her full measure of years in Mars. Since only the perfect Martian women may become mothers of the race, Martia had elected, through some unfitness, to be suppressed.

Finally, Martia leaves Barty Josselin to take up her abode within his child, where she hopes to increase the speck which Barty has at the base of the brain. This speck represents a rudimentary sixth sense which, if cultivated, should grow in men as big as a millet seed;

> And then it will be as big as in a Martian, and the earth will be a very different place, and man of earth greater and even better than the Martian by all the greatness of his ampler, subtler, and more complex brain.[18]

In his autobiography written in 1934, H. G. Wells swings back through the years to make a tardy acknowledgment of his "mental indebtedness" to Grant Allen, and to speak of the bonds of interest that were shared by the two popularizers of science.[19] Like Wells, Allen, twenty years older, had been a science teacher and "full of the new wine of aggressive Darwinism." Like Wells he had never found a footing in the professional scientific world, but had got an infection of that same ferment of biology and socialism that was to work in Wells' blood, and had become, of consequence, a novelist of ideas.

Wells's preoccupation with the future in his novels need not be derived, however, from Allen or works like *The British Barbarians.* Direct contact with evolutionary speculation at his most receptive age, as the philosopher of Utopia himself suggests, played by far the largest part in the matter.[20] He passed through an imaginative adolescence at a time when Weissmann was in the ascendant and the mechanism of evolution was a field for almost irresponsible reconnoitering. Three years of "illuminating and good scientific work" under Huxley at the University of London's Normal School for Science, one year of which was spent in Huxley's class, gave Wells the evolutionary bent which affected the whole of his subsequent writings.

The first ten years of Wells's work in fiction, 1895-1905, he filled with quasi-scientific tales of wonder and with romantic prophecies of the future in the vein of such works as Bulwer Lytton's *The Coming Race,* W. H. Hudson's *The Crystal Age,* and Edward Bellamy's *Looking Backward*—imagined reconstructions of society in some Utopia out of the real world. Later he turned to the composition of sociological and psychological novels of present day life. But he continued to send spinning out at intervals a dozen futuristic creations, vitalized by his steadfast ardent belief that human progress is not only painfully necessary but actually possible. Wells never failed to take the natural sciences with him into his dreams of a new world and criticisms of the existing social order. With the courage of his sanguine imagination he continued to apply scientific methods to make mankind pacific, intelligent, well, and wealthy.

In *A Modern Utopia* (1905)[21] Wells does not paint a picture of a Utopian state, but rather points out the path for reaching Utopia. Like Du Maurier he holds that static perfection is not only impossible but undesirable. Wells' creed of creative evolution, in common with that of Butler, Shaw, and the Vitalists, is the optimistic one that the moulding of our selves and our future lies in our own hands and that if man will only want and try hard enough he can so evolve as to gain complete power over himself and his environment. Wells is impatient with the philosophical evolutionists who spread the idea abroad that the average man is improving by virtue of the same impetus that raised him above the apes; that we can all fold our hands over our paunches in the serene faith that the universe is being managed for us better than we can possibly control it for ourselves. Dame Nature, he insists, following Huxley here, can no longer be looked upon as a kind mother leading her children by her apron strings, but as an indifferent hag, at times inefficient, at times cruel, who has to be bestridden and tamed.

By a miracle of legerdemain Wells projects us into a Utopia of some-when, occupying not some subterranean country or tract "over the range" but a whole planet, a planet which at every point tallies with our own.

The Utopia of a modern dreamer must needs differ in one fundamental aspect from the Nowheres or Utopias men planned before Darwin quickened the thought of the world.[22] Principally, in Wells' New Republic, this involves the presentation not of a finally perfect state of *being,* but a hopefully ascending one of *becoming;* the acceleration of physical and mental evolution by birth-control; the consideration of all measures in terms of the human race as a whole, not of any political or racial subdivisions. Not artificial but natural psychological differences determine his stratification of classes in Utopia. From the upper two classes, the kinetic or intelligentsia, the poietic or creative, he would recruit the governors of Utopia whom he calls the Samurai.

The modern Utopia lays emphasis on the improvement of the race by selective breeding. It was in studying the methods of pigeon breeders that there was suggested to Darwin the conception of natural selection. As soon as the importance of artificial selection in the transformation of species of animals was understood, reflection naturally turned to the human species, and the question arose: How far do men observe, in connection with themselves, those laws of which they make practical application in the case of animals? Darwin presented the question and answer both in his Conclusion to *The Descent of Man:*

> Man scans with scrupulous care the character and pedigree of his horses, cattle and dogs before he matches them; but when he comes to his own marriage he rarely, or never, takes any such care.[23]

It was such a consideration which guided the researches of Francis Galton, Darwin's cousin, on the laws of human inheritance.

Realizing that the species suffers when the "fittest" are not able to perpetuate their type, Wells sought by artificial selection to improve the stock in his planned future. Here as elsewhere he refused to rely on the ways of Providence. (Providence? "Evolution was our blessed word" for it, was the later comment of Mr. Barnstaple speaking of a similar "do nothing" attitude.[24])

The method of nature "red in tooth and claw" of raising the level of the average by killing the weaker and least adapted members is slow and fraught with too much pain. Not by killing the weak,* but by preventing the birth of those who would be bound to suffer and fail do the Utopians propose to provide for the survival of the fittest. Conditions are laid down for limiting parentage, providing that in order to be a parent a man must rate above a certain minimum of physicial and mental capacity, and income. By such an evolutionary process the unfit and undesirable will be eliminated.

Wells is incensed at the attempt of jingoists to seek justification for their patriotic conceit in the doctrines of Darwinism. Confounding race with nationality this modern school of imperialism interprets the conflict among nations as a struggle for existence between superior and inferior types. The races not prospering politically such as the Egyptians, the Chinese, the Hindoos, and all uncivilized people are represented as inferior. Those in the ascendant are assumed to be the best races, whose moral duty it is to exterminate the inferior. Such a distinction

---

* Grant Allen represents the ruthlessness of such a policy in a future state of society where the destiny of the race is considered more important than the feelings of any individual. A clubfooted child is born to a mother who is none too strong. Thinking "first and foremost of the progressive evolution of universal humanity" the governors of the community decree that the child must die. The natures of the phalansteric parents—and the people in general, however, are only half attuned to the high social principles involved. On "Darwin, December 20," the child is executed. The mother, too dies.

Grant Allen, "The Child of the Phalanstery," in *The Backslider* (New York and London, 1901), pp. 315-342.

"has a fine modern biological air ('Survival of the Fittest')." There is a Teutonic "best" race, an Anglo-Saxon "best," a Roman "best." And over and above these is a broader imperialist school embracing the whole white race as the superior type. This is the school that creates slogans like "the white man's burden" and speaks belligerently of "the yellow peril." Exponents of this doctrine look with a resolute, truculent eye to a future in which all the rest of the world will be in subjection to these elect. Wells, whose philosophy took shape in the biological laboratory, will recognize no racial or political boundaries in his modern Utopia. With a larger mind that looks across frontiers he envisages a world-wide synthesis of all cultures and polities and races into one World State.

With the same eager projecting and reforming temper that quickened the spirit of Shelley, with an equally Shelleyan faith in the possibility of refashioning human nature—but in evolution's image, Wells continued to express and exploit a future peculiarly his own. He is accepted as the Mohammed of the scientific era, clad in a prophet's mantle. Years ago his cosmic Koran was complete. But enraptured disciples clamored for new apocalypses. He has lived to have new visions of a world set free, of a world where men are like gods, visions of the shape of things to come. Throughout he has continued to preach his creed that man is at the dawn of change "from life regarded as a system of consequences to life regarded as a system of constructive effort."[25] The view is a non-fatalistic one, opposed to the dogma of the neo-Darwinists who can see no hope for mankind except in fortuitous changes which lie outside his control. To Wells man is a being of intellect who chooses his own path. The story of evolution is not one to inspire dismay in its hearers' hearts but rather determination and illimitable hope.

CHAPTER FOURTEEN

THE COMING RACE AND THE IDEA OF DEGENERATION

That the world was "going to the dogs" was the oft-expressed belief of many who had never even heard of the theory of evolution. But when scientists, of sufficient number and authority, pontificated solemnly on the future dissolution of the world, the seal of validity was given the belief. Some looked foward to a short, quick end—a catastrophic collision with a comet, perhaps. But the long view of development afforded by the theory of evolution had reenforced the suggestion of an infinite future. Even those mildly pessimistic realized that "if for millions of years our globe has taken the upward road, yet sometime the summit will be reached and the downward route be commenced."[1] Consequently, most deteriorationists gloomily anticipated a world in which the vicious tendencies of the present would be enlarged and exaggerated; in which imperialism and individualism and materialism would have driven to extinction all the finer motives in life. That such a civilization could not last, but, worm-eaten with selfishness, would break as a beam snaps, was likewise realized as inevitable. Such decay might come, it was predicted, not alone to a social organization essentially corrupt, but also to one which had reached a plateau of static perfection. This gave rise to a theory of cyclic evolution: successive eras of the development, collapse, and reconstruction of society.

Not all agreed with the assumption contained in most of the black-bordered blue-prints of the future that, come what may, man, whatever his character, would continue to dominate the earth. Invasion from another planet by a race of inhuman creatures was one romantic possibility. The devolution of man into a monstrous caricature of himself through the atrophy of all but his brains and hands, was another, suggested by the accumulation of vestigial relics in man's anatomy. To those who concentrated on the impersonal, fortuitous, and purposeless nature of evolution, anything was possible. Nature gives no guarantees. Our race "may dwindle and fade out altogether before some emboldened animal antagonist, or through some world-wide disease brought to it by rats and dogs and insects . . . who may be destined to be heirs" of us all.[2] From scarped cliff and quarried stone nature lays bare a thousand types that have gone down to extinction. Finally, evolution may be a progress in the wrong direction, "a graceless drift towards a dead end."[3]

The representation of degeneration in the romance of the ideal world follows the pattern set by the Utopian romance. Degeneration is unfolded through the medium either of a strange world or of a future world, peopled by creatures with a far longer evolutionary history than the human. The romance of the strange world exhibits creatures, relatively human, evolved under alien circumstances from a stage originally human or quasi-human. It is intended to represent the ultimate development of man on earth, as much as does the literal romance of the future.

In the representative novels exhibited here, the Society of the Future is poised on the edge of dissolution in a state of stagnant perfection; or about to be overrun by a new species more potent and pitiless in its destructive power than the race of man; or already degenerate, fading into the cold twilight of the world. In the last, man, the microcosm, is viewed in cosmic perspective, on a background of infinite space and eternal time: "an inhabitant of a thin rind on a negligible detached blob of matter belonging to one among millions of stars in one among millions of island-universes."[4] His passing is no more than the flutter of a leaf in the forest.

In a state of stagnation or, at best, static perfection are the races of beings described in romances by Percy Greg and by W. H. Hudson, the one set in Mars, the other, somewhere under the earth. If the race on Mars in Greg's *Across the Zodiac* (1885)[5] is meant to be a picture of the future race of man, progress is more to be dreaded than revered. That the author intended the Martial race to prefigure humanity-to-come is apparent in numerous comparisons, and in the teleological error of supposing that Martians have evolved along the path of human evolution.

The story follows the pattern set down by Jules Verne. What appears at first to be a meteor smashes into an island in the Pacific. Investigation results in the discovery among the debris of human flesh and bone, and a metallic case of strange alloy. Within the case is a book which, deciphered, recounts the story of a journey to Mars in a car made buoyant by apergy, an electric force operating counter to gravity.

The geology of Mars implies the existence of man upon its surface "ten, twenty, or even a hundred fold longer than he is supposed to have enjoyed upon earth."[6] But with this time advantage over society on earth the Martians have developed a mechanically perfect but uninspiring civilization.

The Martians are diminutive in stature, delicate as hot-house plants in constitution. Disease and old age have been done away with, yet the people somehow die of no other apparent cause than being tired of living. But in the absence of disease on Mars the people have lost the power to resist it. Hence, when the Martial wife of the traveler catches an earthly disease while tending roses grown from seeds the traveler brought with him, she dies.

Their delicacy to the contrary, the Martians practice scientific selection. Children are cared for and educated in creches. By euthanasia all deformed and defective children who might be the cause of physical deterioration in the race, are promptly and painlessly deprived of life. But inanition takes its toll; for science and invention have made labor an anomaly. Brutes—squirrel and monkey-like creatures—have been trained to service; and with the regulation of the nature of domestic animals, Mars has become a land of Cockaigne. Unicorns come regularly to be milked at sunset, and if told before-hand come earlier or later to any place pointed out to them.

Maternal affection is discountenanced on the planet as a purely animal instinct, and family affection has become extinct. With these have gone all public spirit and religion, both of which were deemed utterly fatal to the progress of the race. In place of religions Martians have set up the worship of science whose infallibility none may challenge.

The traveler becomes a member of a secret society, the Order of the Star, which connives to overthrow the worship of material science and restore the worship of God. As a member of the order he is vouchsafed a motion-picture view of the evolution of cosmos and of Martial man.

The physical and biological evolution of Mars is presented as strikingly similar to that of the earth. There is a representation first of the formation of the universe according to the nebular hypothesis; then of the evolution of Mars; finally, of the evolution of life, from a tiny globule by slow degrees into higher and more perfect shapes—to man. From which of the animal forms man evolved, the traveler cannot say. "But that no true ape appeared among them, I do distinctly recollect."[7] Nevertheless, he does see in man a spiritual essence not present in any lower form.

Evolution in the Martial sea does not seem to have taken place *pari passu* with evolution on land; in fact it is markedly anachronistic. The adventurer on Mars actually discovers the surviving prototypes of prehistoric monsters.

> Creatures of a type long since supposed to be extinct on earth still haunt the depths of the Martial seas; and one of these—a real sea-serpent of above a hundred feet in length and perhaps eight feet in circumference—had attacked our vessel.[8]

And even a species of the extinct pterodactyl or winged reptile exists, hunted by the Martians, mounted, (like the Kosekin of *A Strange MS.*) on riding birds.

Society has not evolved in a straight line on Mars. Its development involved stages similar to those on earth, through communistic experiments, finally to the development of a single state—a monarchy. But the cycle is not complete; for as the traveler leaves, the secret society is on the verge of transforming the old order.

Cyclic too is the view of the progress of civilization described in W. H. Hudson's *A Crystal Age* (1887).[9] The great cities and the complicated metropolitan customs that they produced have long since been wiped away. The collapse of the old many thousand years before was but a prelude, however. to the new. In the place of the urban civilization is a pastoral one, "a dream and picture of the human race in its forest period."

How this idyllic civilization came into being and where, is shrouded in mystery. Smith, a young Englishman, on a botanizing expedition in the mountains is suddenly knocked unconscious when the ground gives way under him. When he awakes his body is entangled by rootlets and covered with mold. The land around him is unfamiliar. He comes upon a strange people who, though they speak English, have never heard of Gladstone, Napoleon, Homer, Shakespeare, Adam or Eve. They take him with them, and Smith becomes one of a family dwelling in an ancient and beautiful mansion. Dimly he becomes aware that many thousands of years before, man tried to gain absolute dominion over nature, and thereby invited catastrophe. It came, carrying with it to destruction all the people of the earth save a few who were men of humble mind and lived apart. These men peopled the earth and founded their placid civilization.

The purest matriarchal type is shown in the dream of a Crystal Age wherein everything centers around the mother worship of the secluded and enthroned matron-head of the family. The household is the social unit; and labor and brotherly love the core of their existence. The rural tasks are far from taxing, for the horses and dogs of the Crystal Age have evolved a degree of intelligence which our common breeds are far from possessing. The horses come willingly from their unfenced pastures and all but harness themselves to the plow; and the dog teaches Smith when to leave off working the animals.

Free from the strange disease of modern life, the weariness, the fever, and the fret of our age, the Crystal-lites appear with the "passionless, everlasting calm of beings who have for ever outlived, and left as immeasurably far behind as the instincts of the wolf and the ape" the emotions of ordinary man. For all except the house mother—a sort of queen bee—who perpetuates the family in each house, sex is a matter of physical appearance only. The Crystallites are content with a vegetable love: a "chill moonlight felicity" is all that remains of human passion.

By this means they practice scientific selection. Smith learns with the deepest despair that his passion for Yoletta can never be reciprocated by his beloved. Only those women may marry who are especially endowed by nature to become mothers. The housemother has a remedy for his grief—a phial of liquid labelled "Drink of me, and for the old life there shall be a new life." He drinks the ice-cold, pale-yellow liquid in the belief that it will make him as free from passion as his housemates; and he is not deceived; for—he dies.

By careful selection, Hudson implied, the Crystallites have achieved the ultimate in man's evolution, a race, living in social harmony, blissfully ignorant of the storm and stress which once tore man asunder. But the poet-naturalist realized later, in this idea of an "unchanging peaceful present" with the stamp of everlastingness upon it, was implicit a contradiction of natural selection. Viewing his work from the perspective of thirty years, Hudson wrote in the preface to the 1906 edition:

> Alas that in this case the wish cannot induce belief. For now I remember another thing which Nature said—that earthly excellence can come in no way but one, and the ending of passion is the beginning of decay.[10]

Only a struggle for existence, Hudson knew, can provide for the survival of the fittest.

The picture of the Martians in H. G. Wells's *The War of the Worlds* (1896)[11] in no way resembles that of the physically human, spiritually decadent Martians of Percy Greg's *Across the Zodiac* or the fur-skinned, spiritually beneficent Martians of George Du Maurier's *The Martian.* And though Hudson's Crystallites and Wells' Martians are both devoid of human emotions, the resemblance ends right there. More of a scientist than his predecessors, Wells set himself the task of conjuring an utterly alien form of life that nevertheless would accord with the principles of biology. "The chances against anything *man*-like on Mars are a million to one," the astronomer Ogilvy remarks in the story.[12] Wells developed in one direction the theory of human evolution expressed in "The Man of the Year Million," an article which he had written for the *Pall Mall Budget.*[13] The resultant man of Mars was a grotesque monster, the apotheosis of pure intellect.

The War of the Worlds is a story of the subjugation of the human race by an invading horde of Martians. Ten huge cylinders shot from Mars crash onto the English down. From them emerge oily monsters, jelly-like organisms all brain and tentacles, equipped with armament against which no human defence can avail. Mounted on tripods that travel with the speed of express trains they level the countryside with their heat rays, annihilate all who stand against them. To the demons from Mars, human men and women are simply "edible ants" to be exterminated in the unsparing way of the biological war among species. Impervious to the guns and cannon, the Martians undertake to subdue the world and are on the point of succeeding in their conquest when a strange ally comes to man's aid: the disease bacteria. Since the Martians have long eradicated disease, they have lost their resistance to microbes and hence are altogether defenceless against them. They sicken and die leaving the world free but without "that serene confidence in the future which is the most fruitful source of decadence."

The picture of man's successors, the "superior" creatures of Mars, is without doubt a depressing one. Wells could not have gone much further in his selection of revolting details for the Martian portrait. The evolutionist had chosen an ape-like progenitor for mankind; now Wells had selected for man's successor a disembodied brain—an armless, legless body with a chinless, noseless face.* But Wells was as much a scientist as a romancer; his nightmarish, vampire-like creatures who took the fresh living blood of other creatures and injected it into their own

---

* In a short story, "The Empire of the Ants," (1898), Wells advanced the romantic thesis that man may sooner or later be supplanted in his dominion over the earth by ants.

"In a few thousand years men had emerged from barbarism to a stage of civilization that made them feel lords of the future and masters of the earth! But what was to prevent the ants from evolving also?"

H. G. Wells, "The Empire of the Ants" (in *The Works of H. G. Wells,* Atlantic edition, vol. X), London, 1925, p. 499.

veins were the product not alone of a fertile imagination but of applied evolution. As a scientist Wells realized that such organs as hair, external nose, teeth, ears, and chin were no longer essential parts of the human being and that the tendency of natural selection would lie in the direction of their steady diminution through the ages. A steady expansion of the brain case could be taken for granted. Aside from the brain only one other part of the body had a strong chance of survival—the hand.[14]

The Martians exhibit just such a suppression of the animal side of the organism by the intelligence. They are not bodies, really, but huge round heads. Their hands have been transformed into tentacles—sixteen of them—grouped around the mouth. The rest of their anatomy consists of enormous nerves ramifying to the eyes, ear, and tactile tentacles. Everything makes for efficiency of action, economy of effort. All the complex apparatus of digestion which bulks so large in our bodies does not exist in the Martians. The injection of blood for food is a tremendous physiological advantage over the human eating and digestive process. The Martians are sexless; they reproduce by budding, like the young of coral. Their limbs are mechanical devices—an improvement for which Wells was indebted to Samuel Butler's suggestion in *Erewhon,* that a mechanical device is only an added limb. The people of Mars have become practically mere brains, wearing different mechanical bodies according to their needs.

Wells does not permit the human application of all this anatomical science to be lost. Again and again he points his evolutionary moral. "The Martians may be descended from human beings not unlike ourselves, by a gradual development of brain and hands . . . at the expense of the rest of the body."[15] And again: ". . . we men . . . are just in the beginning of the evolution that the Martians have worked out."[16]

Darwin's evolutionary thesis, natural selection, is Wells' explanation both for the coming and the extinction of the Martians. Struggling for their existence on an ever-cooling planet, the Martians realized the necessity of carrying on an imperialistic war sunwards if they were to escape from destruction. Cold, remorseless intelligences unswayed by emotion, the Martians no more hesitated to trample men underfoot than we do an insect.* With the objectivity of the scientist Wells draws the analogy of man's utter and ruthless destruction of inferior races to gain a place in the sun. Finally, natural selection, enlisted for the moment on the side of the earthlings, turns the tide against the invaders from the red star. The germs of disease have taken toll of humanity since the beginning of time. By virtue of natural selection man has developed resisting power and even immunity against them. But if ever the Martians had that power to repel disease, they lost it, and with it their chance to overrun the earth.

Mankind might regard with less despair the achievement of a stagnant perfection as in *Across the Zodiac* or *The Crystal Age* than the ultimate succession of Martian-like humans, or ants or Dimensionists. The choice of static civilization, however, would be the choice only of the lesser of two evils. In the earthly paradises, perfect but monotonous, which Greg and Hudson depicted, not even the practice of artificial selection served to arrest imminent decay. Rather, since the race was being scientifically groomed for the very purposes which were contributing to the decay, such eugenics was merely hastening the end.

---

* Joseph Conrad depicts an equally ruthless successor to man in an extravaganza, *The Inheritors* (1901). Conrad's candidate is a race from the Fourth Dimension. Callous to pain, they have no more regard for weakness, suffering, and death than if they were invulnerable or immortal. Like the Martians, the Dimensionists are being crowded out of their native habitat, and threaten to invade the earth and "to devour like locusts."

Joseph Conrad and Ford M. Hueffer, *The Inheritors* (Garden City, New York, 1926).

Not artificial but natural selection brings about the disintegration of man and society in H. G. Wells' *The Time Machine* (1895).[17] Wells assumes that an exaggeration of contemporary conditions has split the human species along the lines of social stratification, into Haves and Havenots. The Havenots have been driven underground and by natural selection have changed physiologically in adaptation to their subterranean environment. The Haves, in full possession of the upper regions, have decayed to a mere beautiful futility.

The curtain is rung up on this decadent civilization in the year 802,701, reached through the marvelous device of the Time-Machine. This, a simple mechanism of crystalline bars and levers which can nullify the "gravitational" pull of time, has sent the Time-Traveler hurtling through the future, night following day "like the flapping of a black wing." His journey through time takes him first to the degenerate age of the Eloi and Morlocks eight hundred millenniums hence; then on through the countless cycles yet to be before the coming of the final disintegration, to the senescence of the earth thirty million years forward. He returns to tell the story of his experiences to the assembled dinner-guests, with no memento but the crushed blossoms of a strange flower.

The Time Traveller had looked forward to the discovery of a magnificent race of intellectual beings ahead of our own. Instead, he came upon the remnants of a Golden Age, long since past its zenith and now fallen on sere and yellow days. Mankind had not remained one species but had differentiated into two distinct races: The Eloi and the Morlocks. The diminutive, childish Eloi dwell on the surface and spend each sunlit day in idle, graceful pastime. But at night the Morlocks, the white horrible ghoul-like creatures who dwell underground, come out to rule, and feed like cannibals upon the fragile creatures of the Upper World.

Evolution is the driving force which has swept mankind on through the rapids to security in a social paradise, there to rot with the need for vigor and effort gone. Wells was merely suggesting a logical conclusion to the industrial system of today. The rulers of mankind had set themselves steadfastly to the task of achieving a balanced society with security and permanency as watchwords. The masses were little by little chained to their machines below the earth, where they were forced to adapt themselves to the conditions of subterranean life—or perish; the classes, having triumphed over nature and their fellow-man by driving the social and economic questions underground, were free to pursue pleasure and comfort and beauty. Physiologically this differentiation took its toll on the human creatures of the Lower and Upper Worlds. The Workers, without sunlight, burrowing and tending machinery and providing for those above, have developed into ape-like creatures with that bleached look common in most animals that live largely in the dark, and whose eyes have the capacity of nocturnal things for reflecting light. By idleness and futility induced by their too-perfect security the Upper-world man has become a light-headed, puny, and helpless creature: for "an animal perfectly in harmony with its environment is a perfect mechanism"—without intelligence or the need for it.

But that perfect state lacked permanence. In the course of time some failure of the food supply for the carnivorous Morlocks drove them to seek for it elsewhere. Inevitably, since "horses, cattle, sheep, dogs had followed the Icthyosaurus into extinction" they turned to cannibalism with the feeble Eloi as their prey. The old order was reversed:

> The two species that had resulted from the evolution of man were sliding down towards . . . an altogether new relationship. . . . Ages ago, thousands of generations ago, man had thrust his brother out of the ease and the sunshine. And now that brother was coming back—changed.[18]

Back in his Time Machine, the Time-Traveller pulls the lever over and wings his way into futurity, the thousands hand of the dial recording years sweeping round as fast as the seconds hand of a watch.

> I cannot convey the sense of abominable desolation that hung over the world. The red eastern sky, the northward blackness, the salt Dead Sea, the stony beach crawling with these foul, slow-stirring monsters, the uniform poisonous-looking green of the lichenous plants, the thin air that hurt one's lungs. . . . I moved on a hundred years, and there was the same red sun—a little larger, a little dulled—the same dying sea, the same chill air, and the same crowd of earthly crustacea creeping in and out among the green weed and red rocks. And in the westward sky I saw a curved pale line like a vast new moon.[19]

And so on he travels through the senile years of an earth hovering over the abyss of space before its final plunge into the sun millions and millions of years hence. And then he comes back only to set out once more backward through time—never more to return.

For the world as for man Wells saw and dramatized the possibility of "a graceless drift towards a dead end." There were others in his day who looked into the future with the same prophetic mood that Wells chose to exploit. But none could coin so well that curious amalgam of scientific knowledge and riotous romance that he himself has compared to the "monstrous imaginings of children." None could match, too, his cosmic grasp of space-time. Though his work is a milestone in the history of evolutionary romance there were few with the grounding in science and the vivid and fluent imagination to follow him into the future.

CONCLUSION

The period from 1860-1910 saw the rise and decline of evolution as a literary theme. Its incidence greatest in the 'seventies and 'eighties in the form of controversial fiction, and from 1885 on in the form of scientific romances, the evolutionary motif did not disappear after 1910, the concluding year of this study. Rather, it faded merely in the glare of more contemporaneous concerns.

What the themes of class struggle and war are to the literature of the present day, the theme of religion's struggle with science was to the Victorian age. What the scientific inventions and discoveries of today—relativity, cosmic rays, atom-smashers—are contributing to the imaginative literature of today, the theories of evolution and natural selection alone contributed to the imaginative literature of the Victorian era. Other problems and interests held the close attention of the age—political reform, social evils, colonial policy, imperialism—but none had such wide ramifications in the field of thought, and few such widespread reflection in the literature of the time.

The doctrine of natural selection which Darwin conceived as a purely biological theory, was soon adapted to other ends.* Having begun in an economic formula—that of Malthus—it was only natural that Darwinism should return to that field.

---

* Cf. Alphonse Daudet's play *La Lutte pour la Vie* (1890), where Darwinism is used by the villains to justify robbery and murder. Daudet explains in the preface:

"Non, j'ai seulement voulu mettre a la scène quelques specimens de cette race nouvelle de petits féroces a qui la formule Darwinisme de "la lutte pour la vie" sert de prétexte et d'excuse en toutes sortes de vilenies et d'infamies."

Alphonse Daudet, "La Lutte pour la Vie" (in *Oeuvres Completes de Alphonse Daudet,* Tome 17), Paris, 1901, Theatre IV, p. i.

Six weeks after the *Origin* Darwin told Lyell: "I have received, in a Manchester newspaper, rather a good squib, showing that I have proved "might is right," and therefore that Napoleon is right, and every cheating tradesman is also right."

Life and Letters of Charles Darwin, II, 56-57.

Struggle for existence in an economic sense is illustrated in George Gissing's *New Grub Street,* and in Samuel Butler's *The Way of All Flesh;* in the one case leading to failure, in the other to a successful adaptation to environment. The ruthless individualist welcomed the Darwinian doctrine as a scientific repudiation of the Christian tenet that "the meek shall inherit the earth." Into such a specious apologue for the rule of might, Mary St. Leger Harrison rips fiercely in *Colonel Enderby's Wife.* In the same mood is Bulwer-Lytton's condemnation in *Kenelm Chillingly* of those "big fishes" who cite universal law in order to justify their heartless destruction of "little fishes." *That Very Mab* by May Kendall scornfully acclaims the manufacturer who considers himself the fittest, through his having grown rich on the proceeds from the sale of shoddy merchandise. Recognizing that the more recent development of the human species has been not physical but mechanical, in the form of "machinate" extensions of personality, Samuel Butler concludes in *Erewhon* that men have had good cause to revise their conceptions of the fittest. These additional "limbs" are likely to be costly; accordingly the right differentiation of civilized man is not by race or character or physical prowess, but by purse.* Such a "limb," too, is the *vril* or electrical power which the Vril-ya of Edward Bulwer-Lytton's *The Coming Race* have evolved; here the evolution of the mechanical (as well as the physical) has achieved an instrumental utopia by relieving man of the burden of drudgery.

That the accumulation of wealth or power by individuals will affect the future development of the race is the sociological application H. G. Wells makes of evolution in *The Time Machine.* Wells there supposes that the division

---

* Professor Dredge in Oswald Crawfurd's *The World We Live In,* (1884) echoes Butler when he describes the doctrine that wealth is the test of merit as the theory of the "survival of the richest."

Oswald Crawfurd, *The World We Live In* (London, 1884), p. 81.

of society into two distinct classes, the workers and the leisured, may end in so rigid a system of caste that far away down the ages the two orders will become distinct and separate. In *The War of the Worlds* Wells suggests a political turn to the doctrine of the survival of the fittest. He compares the Martian attempt to conquer the earth to the attempts of imperialistic powers to subdue so-called inferior races. The struggle for existence on Mars likewise led the Martians to overrun the borders of their own planet. Again, in *A Modern Utopia,* Wells returns to the political application of Darwinism in a scathing attack on imperialists and nationalists. These, Wells rasps, assume the existence of a best race—always their own—and conceive it their function to exterminate the inferior races, that the fittest only may remain. Like Wells, Olive Schreiner in *From Man to Man* roundly condemns this pseudo-scientific classification of races.

Darwin's warning on man's indifference to questions of eugenics in marriage, and the evident need of direction if human nature is to profit by the lessons of evolution, impelled writers to consider the question of selective breeding. George Meredith in *The Egoist* looks hopefully for race improvement to Darwin's law of sexual selection applied to humans. So too does Grant Allen who, in *A Splendid Sin,* sees a progressive discrimination in the selection of a mate, the higher the scale of life.

Authors of the Wellsian prophetic habit of mind almost unanimously endowed their Utopias with some system of eugenics. In this regard they are all like Aunt Phoebe of H. G. Wells' *Joan and Peter* (1918), who considered race improvement a panacea:

> "I believe in Sir Francis Galton," Aunt Phoebe remarked. "Eugenics. . . . Men ought to be bred like horses. No marriage or any nonsense of that kind. Just a simple scientific blending of points. Then Everything would be different."[1]

There is no agreement on the measures to be taken for improving the breed. The feeling seems to be general, however, despite Grant Allen and George Meredith's optimistic view, that the matter cannot be left to the law of natural selection, unaided by man: natural selection is too slow.* In Robert Ellis Dudgeon's *Colymbia,* the drastic practice of infanticide does away with the unfit; and euthanasia is relied on in Percy Greg's *Across the Zodiac* to achieve the same purpose. But the inhumanity of such a method was deemed as cruel as nature's own. Not by killing off the weak and undesirable but by preventing them from procreating, it was suggested, should the human stock be bettered. In W. H. Hudson's *The Crystal Age* only those women may marry who are especially endowed by nature to become mothers. The Martians in George Du Maurier's novel of that name consider unfitness of any sort, physical or mental or moral, sufficient grounds for preventing the begetting of children; and Du Maurier adds, "we shall have to do so here some day, or else we shall degenerate and become extinct; or even worse."[2] Minimum requirements, physical, mental and economic are set up in H. G. Wells's *A Modern Utopia* as prerequisite to bringing children into this world. Finally, Olive Schreiner in *From Man to Man* rehearses all the arguments for regulating procreation, but ends her discussion with a straddling of the issue. Though she is whole-heartedly opposed to leaving the matter up to natural selection, Mrs. Schreiner predicates the conviction that there can never be general agreement on what constitutes superiority, or on who shall be the arbiters of such a momentous question.

---

* In Richard Whiteing's *The Yellow Van* (1903), gazing at the peasant woman Mrs. Patch, the hero, Arthur wonders "whether the evolutionary system might not possibly be brought under new management for the speedier elevation of the race. It might otherwise take thousands of years to give Mrs. Patch a lift toward the skies."

Richard Whiteing, *The Yellow Van* (New York, 1903), p. 257.

The catch phrases "struggle for existence," "survival of the fittest," "missing link," occurring time and again in theme and dialogue, and the frequent representations of prehistoric monsters like the plesiosaurus and the icthyosaurus, gave the Victorians a sense of familiarity with the doctrine of evolution and the doctrine of natural selection. Even across the channel Alphonse Daudet chose to translate his novel *L'Immortel* into a play *La Lutte pour la Vie* and to designate his villains by the unique cognomen "struggle for lifeurs." So large a number of Victorian writers were engaged in the task of exploiting the theme of evolution, that it was inevitable that a doctrine originally confined to the attention of scholars should filter down to the general reading public. It was equally inevitable that the public's uninformed sympathies for one side or the other of the religio-scientific conflict, would be sought through the medium of special pleading in the novel. Much more aggressive by far were the protagonists for religion in the literary battle that echoed the one taking place in pulpit and lecture-hall. "Materialist" (like "Communist" today) was used again and again as a weapon of offense and defence in the guerilla warfare conducted against the leaders of science. Our tears are called for in such a work as Marie Corelli's *The Mighty Atom* for the little boy whose God was torn from him by the brutish father who believed only in "atoms." An atheistic scientist in the Victorian didactic novel, indeed, especially in one by a lady novelist, figures conventionally as "the shocking example" to be killed in the last chapter of the last volume under particularly horrible circumstances—or else converted. Such novels as Catherine Liddell's *Margaret,* Julia Kavanagh's *John Dorrien,* Eustace Claire Grenville Murray's *That Artful Vicar,* and *Paul Nugent, Materialist* by Helen F. Hetherington Gullifer and the Reverend H. Darwin Burton, are representative.

The theme of May Kendall's *That Very Mab* that under the eye of science all illusion vanishes, and that when science came in the window, art and imagination flew out the door, was hardly realized in the flood of imaginative fiction that swept in with Darwinism. The demands of artistic imagination were satisfied with the creation of the genealogical novel exemplified in *The Way of All Flesh* in which scientific heredity determines the nature of the characters. The romantic imagination which had been at work creating Gothic castles and supernatural horrors now gave way to the scientific imagination which delved into the past and probed into the future, peopled the planets and the hollows of the earth—in accordance with evolutionary principle. Some novels exhibit a coalescing of the modern with the antiquated forms of fiction. All of the inhuman horror of the Gothic tale informs such a work as Wells's *The Island of Dr. Moreau;* but it is more in the laboratory tradition of Mary Shelley's *Frankenstein.* George Du Maurier's *The Martian* follows the Gothic in its infusion of the supernatural element; but this has been curiously blended with the Utopian romance.

Artistic or scientific imagination does not distinguish all fictional representations of the theory of evolution. In many pieces of literature turning on an evolutionary theme little or no attempt was made to fuse science with art to achieve an imaginative synthesis. Rather, these carried to an extreme the practice instituted by Victorian novelists of padding the story with dinner-table discussions of contemporary problems, to eke out the conventional three-volume novel. There is not enough story to pack into a thimble in these treatise-novels. Thomas De Quincey's well-known distinction between the literature of power and the literature of knowledge here offers a ready-made explanation. The function of the literature of power is to move; the purely imaginative fiction had little other purpose.

The function of the literature of knowledge, however, is to teach; and this function, in a society whose most serious reading is generally the novel or magazine, many writers considered paramount. They were ready to make the slight concessions to story merely to keep the work out of the non-fiction class.

The novels of W. H. Mallock, Samuel Butler, Olive Schreiner, and H. G. Wells serve this second or didactic function. Only by a very loose definition of terms could such a *roman à thèse* as Mallock's *The New Republic* or his *The Veil of the Temple* be called a novel. The characters in these works who delight in lengthy dialogue, disquisitions, and even sermons, are mere mouthpieces of the author's evolutionary views. Samuel Butler's *Erewhon* contains in "The Book of the Machines" chapters, a monumental treatise on the evolution of machines which, though in keeping with the satire, is hardly an integral part of the story. In Olive Schreiner's *From Man to Man,* the musings of Rebekah on the function of love in the survival of species are almost without exception interpolations without relation to the plot or story. In this regard H. G. Wells has been perhaps the chief offender. In such a work as *A Modern Utopia* Wells's main interest is sociological; and the romantic science is merely an ill-fitting patch upon the presentation of the serious questions occupying the writer's mind. The resulting combination is neither fish nor fowl, neither romance nor novel.

Between the two divisions of literature spoken of by De Quincey lies a kind of writing which occupies a considerable space in the Victorian era. It tries at once to enlighten the intellect and to quicken and refine the sensibility. In such a category could be placed novels like Samuel Butler's *The Way of All Flesh,* the majority of the Utopian romances and romances of the future like Edward Bulwer-Lytton's *The Coming Race,* George Du

Maurier's *The Martian,* and H. G. Wells's *The Time Machine.*

Evolutionary literature since 1910 has taken almost exclusively the form of romance, following lines laid down before 1910. Such romances have had to compete, in America at least, with comic strips and radio programs exploiting the far-flung future and man's caveman past.

Following the example of Jules Verne, of J. P. Webster's *Oracle of Baal,* and of J. Compton-Rickett's *Quickening of Caliban,* Arthur Conan Doyle described in *The Lost World* (1912) the discovery in the South American jungle of living fossils of the missing link and of prehistoric monsters. Continuing to revel like a medieval mystic in dreams of things to come, H. G. Wells turned to the fourth dimension in *Men Like Gods* (1913), to set forth his maturing ideas of the Man of the Future. In *Mr. Blettsworthy on Rampole Island* (1928) Wells followed Doyle to the discovery of a land peopled with prehistoric monsters; but these are simply pegs on which the author hangs his usual sociological moral. In 1937, in *Star-Begotten,* he renovated his fantasy of four decades before, *The War of the Worlds,* and pictured Mars once again bombarding the earth—this time with cosmic rays, which, creating mutations, accelerate the evolution of the race. Sidney Fowler Wright is by general acclaim the disciple of Wells who most nearly approximates the master's cosmic touch. Wright's *The World Below* (1929), in the direct line of Wells's *The Time Machine,* imagined a world half a million years hence when the human race consists of giant Dwellers, chiefly male and living mostly underground; and Amphibians, chiefly female, living mostly in the sea. Another of Wells's disciples, Aldous Huxley, presented in his satiric *Brave New World* one of the most bitter indictments of modern civilization and its tendency to standardization. Taking suggestions both from Wells's *The Time Machine* and from his *The First Men in the Moon,* Huxley portrayed the world of the twenty-fifth century whose fundamental theme is the regulation of the race through scientific breeding of human beings in laboratory-factories.

Novels like anthropologist Robert Briffault's *Europa* (1935) may call to mind the throes and perplexities of religious doubt in the years following Darwin, but Briffault's references are mere echoes from a distant scene. To the present day novelist the controversy over evolution is one with "Bertha Broadfoot, Beatrice, Alice, Ermengarde" and all the snows of yesteryear. Only a romanticizing of the theory can have its appeal to the modern reader—a lifting of the veil into a future laid out along the lines of evolution. This glimpse of the future will always have, as it always has had, its appeal to man; but the future of men who know themselves to be super-apes is bound to be different from that of men who believe themselves to be fallen angels.

"Whether your Creator shaped you with fingers, or tools, as a sculptor would a lump of clay, or gradually raised you to manhood through a series of inferior forms, is only of moment to you in this respect—that in the one case, you cannot expect your children to be nobler creatures than you are yourselves—in the other, every act and thought of your present life may be hastening the advent of a race which will look back to you, their fathers (and you ought at least to have attained the dignity of desiring that it may be so,) with incredulous disdain."[3]

REFERENCE NOTES

INTRODUCTION *and* CHAPTER ONE

SURVEY OF EVOLUTIONARY THOUGHT TO 1859

1. H. G. Wells, *Joan and Peter* (New York, 1926), p. 34.
2. Samuel Butler, *Evolution Old and New* (London, 1879), pp. 78-172.
3. Quoted in Donald Culross Peattie, *Green Laurels* (New York, 1936), p. 66.
4. Erasmus Darwin, *Zoonomia* (Dublin, 1800), I, section XXXIX, chap. 4, p. 569.
5. *Ibid.*, p. 572.
6. Erasmus Darwin, *Phytologia* (London, 1800), section XIX, chap. 7, p. 556.
7. Erasmus Darwin, *Zoonomia*, p. 570.
8. L. Rice-Orley, ed., *Poetry of the Anti Jacobin* (Oxford, 1924), pp. 58ff. and pp. 85ff.
9. Samuel T. Coleridge, "Biographia Literaria" (in *The Complete Works of Samuel Taylor Coleridge*, ed. W. G. T. Shedd, Vol. III), New York, 1884, p. 155.
10. Samuel Butler, *Evolution Old and New*, p. 195.
11. Frances Darwin, ed., *The Life and Letters of Charles Darwin* (New York, 1896), p. 34.
12. Francis Darwin, ed., *op. cit.*, II, 198-199.
13. Charles Darwin, *Charles Darwin's Diary of the Voyage of the H.M.S. Beagle*, ed. Nora Barlow (Cambridge, 1933), p. xiii.
14. Francis Darwin, ed., *op. cit.*, I, 68.
15. John T. Merz, *History of European Thought in the 19 Century* (Edinburgh, 1904-12), II, 318.
16. Leonard Huxley, *Life and Letters of Sir Joseph Dalton Hooker* (New York, 1918), I, 489.
17. Frances Darwin, ed., *op. cit.*, I, 302.

CHAPTER TWO

EVOLUTIONARY THOUGHT IN THE NOVEL BEFORE 1859

1. William Knight, *Lord Monboddo and Some of His Contemporaries* (London, 1900), pp. 19-20.
2. A. Martin Freeman, *Thomas Love Peacock* (New York, 1911), pp. 41-42.
3. Thomas Love Peacock, *Melincourt* (London, 1896).
4. Lord Monboddo, *Ancient Metaphysics*, quoted in Freeman, *op. cit.*, p. 43.

5. James Boswell, *Journal of a Tour to the Hebrides with Samuel Johnson,* ed. F. A. Pottle and C. A. Bennett (New York, 1936), p. 80.
6. *Ibid.,* p. 27.
7. Washington Irving, *History of New York* (Phila., 1836), I, 58.
8. Thomas H. Huxley, *Man's Place in Nature and Other Essays* (New York, 1927), pp. 12-13.
9. Thomas Love Peacock, *Headlong Hall* (London, 1896).
10. *Ibid.,* pp. 34-35.
11. Peacock, *op. cit.,* p. 50.
12. Lord Byron, "Letters and Journals," V, in *The Works of Lord Byron,* (London, 1922), p. 317.
13. Thomas Love Peacock, *Nightmare Abbey* (London, 1896).
14. *Ibid.,* p. 183.
15. Francis Darwin, ed., *op. cit.,* I, 314-318.
16. Elizabeth Haldane, *Mrs. Gaskell and Her Friends* (London, 1930), p. 277.
17. George E. Buckle, and William F. Monypenny, *The Life of Benjamin Disraeli, Earl of Beaconsfield* (New York, 1929), I, 852.
18. Earl of Beaconsfield, "Tancred" (in *Novels and Tales* by the Earl of Beaconsfield, Vol. IX), London, 1900.
19. *Ibid.,* pp. 16-17.
20. Robert Hunt, *Panthea, the Spirit of Nature* (London, 1849).
21. *Ibid.,* p. 81.
22. *Ibid.,* p. 85.
23. *Ibid.,* p. 87.
24. *Ibid.,* p. 92.
25. *Ibid.,* p. 342.
26. Charles Kingsley, *Charles Kingsley: His Letters and Memories of His Life,* ed. by his wife (New York, 1877), p. 203.
27. Charles Kingsley, *Alton Locke* (New York, 1861).
28. Margaret F. Thorp, *Charles Kingsley* (Princeton, 1937), p. 78.
29. Kingsley, *op. cit.,* pp. 320-321.
30. *Ibid.,* p. 157.
31. William H. Smith, *Thorndale: or, the Conflict of Opinions* (Edinburgh, 1858).
32. *Ibid.,* p. 313.
33. *Ibid.,* p. 540.
34. Edward Bulwer, Lord Lytton, *What Will He Do With It?* (New York, n.d.).
35. *Ibid.,* p. 160.
36. *Ibid.,* p. 157.
37. Francis Darwin, ed., *op. cit.,* I, 66.

CHAPTER THREE

EVOLUTION AND DARWINISM FROM 1859

1. Charles Darwin to T. H. Huxley, December 2, 1860, in *Life and Letters,* ed. Francis Darwin, II, 147.
2. Charles Darwin, "Introduction," *Origin of Species* (New York, n.d.), p. 1.
3. Thomas H. Huxley, *Man's Place in Nature,* p. 150.
4. Leonard Huxley, *Life and Letters of Thomas Henry Huxley* (New York, 1901), I, 217.
5. Charles Darwin, *Descent of Man and Selection in Relation to Sex* (New York, n.d.), p. 696.
6. Charles Darwin, *op. cit.,* p. 5.
7. Robert Briffault, "Evolution of Human Species" (in *The Making of Man,* ed. by V. F. Calverton), New York, Modern Library, n.d., pp. 761-770.

CHAPTER FOUR

THEOLOGY VERSUS EVOLUTION AND DARWINISM

1. Sir Esme Wingfield-Stratford, *Those Earnest Victorians* (New York, 1930), p. 324.
2. The account of the Huxley-Wilberforce combat is taken from: John W. Cunliffe, *Leaders of the Victorian Revolution* (New York, 1934), pp. 167-169; and Leonard Huxley, *Life and Letters of Thomas Huxley,* I, 197-199.
3. Leonard Huxley, *op. cit.,* I, 197.
4. Francis Darwin, *op. cit.,* II, 115.
5. Andrew D. White, *A History of the Warfare of Science With Theology* (New York, 1930), I, 71-72.
6. Leonard Huxley, *op. cit.,* I, 297.
7. George E. Buckle and William F. Monypenny, *The Life of Benjamin Disraeli,* II, 108.
8. Janet E. Courtney, *Freethinkers of the Nineteenth Century* (London, 1920), p. 154.
9. Andrew D. White, *op. cit.,* I, 76.
10. D. A. Wilson and David W. MacArthur, *Carlyle in Old Age* (London, 1934), p. 328.

CHAPTER FIVE

SATIRES ON EVOLUTION AND THE "MONKEY THEORY"

1. F. G. Walpole, *Lord Floysham* (London, 1886), II, 21.
2. See *ante,* chap. II.
3. Earl of Beaconsfield, General Preface, "Lothair" (in *Novels and Tales* by the Earl of Beaconsfield, Vol. X), London, 1900, p. xv.

4. General Preface, *Lothair*, p. xvii.
5. *Ibid.*, p. xv.
6. *Lothair*, p. 150.
7. *Ibid.*, p. 411.
8. *Ibid.*, pp. 197-198.
9. Charles Reade, *Put Yourself in His Place* (London, 1922).
10. *Ibid.*, p. 155.
11. *Ibid.*
12. Catherine Liddell, *Margaret* (New York, n.d.).
13. *Ibid.*, p. 161.
14. Wilkie Collins, "The Law and the Lady" (in *The Works of Wilkie Collins*, Vol. V), New York, 1875.
15. *Ibid.*, p. 433-434.
16. Wilkie Collins, "Heart and Science" (in *The Works of Wilkie Collins*, Vol. XXV), New York, 1875, p. 538.
17. R. D. Blackmore, *The Remarkable History of Tommy Upmore* (New York, 1884), p. 3.
18. Charles Kingsley, *Water Babies* (Philadelphia, 1899).
19. Clarence Ayres, *Huxley* (New York, 1932), pp. 68-69.
20. Charles Kingsley, *His Letters and Memories*, p. 322.
21. *Water Babies*, p. 117.
22. Edward Jenkins, *Lord Bantam* (New York, 1872).
23. *Ibid.*, p. 204.
24. Edward Bulwer, Lord Lytton, *Kenelm Chillingly* (New York, n.d.).
25. *Ibid.*, p. 36.
26. *Ibid.*, p. 152.
27. *Ibid.*, p. 171.
28. *Ibid.*, pp. 322-323.
29. *Ibid.*, p. 322.
30. Mortimer Collins, *Transmigration* (London, 1874).
31. *Ibid.*, I, 66.
32. *Ibid.*, I, 65.
33. Julia Kavanagh, *John Dorrien* (New York, 1875).
34. *Ibid.*, p. 350.
35. *Tommy Upmore*, p. 3.
36. *Ibid.*, p. 8.
37. Marie Corelli, *Ardath* (New York, n.d.).
38. *Ibid.*, p. 440.
39. Hetherington, H. F. and H. D. Burton, *Paul Nugent, Materialist* (New York, n.d.).
40. *Ibid.*, p. 264.
41. W. H. Mallock, *The New Paul and Virginia* (London, 1890).

42. W. H. Mallock, *Memoirs of Life and Literature* (New York and London, 1920), p. 90.
43. *The New Paul and Virginia*, p. 18.
44. *Ibid.*, pp. 117-118.
45. *Lord Floysham*, II, 42.
46. John Davidson, *Earl Lavender* (London, 1895).
47. *Ibid.*, p. 192.

CHAPTER SIX

SATIRES ON DARWINISM

1. William Alfred Eddy, "Introduction," *Erewhon* by Samuel Butler (New York, 1930), pp. xiv-xv.
2. Samuel Butler, *Unconscious Memory* (New York, 1920), p. 11.
3. Henry F. Jones, *Samuel Butler—A Memoir* (London, 1920), I, 100.
4. Samuel Butler, *A First Year in Canterbury Settlement*, ed. R. A. Streatfeild (New York, 1915), pp. 152-154.
5. *Ibid.*, pp. 152-153.
6. Eddy, "Introduction," *Erewhon*, pp. 1-2.
7. *Erewhon*, p. 266.
8. H. F. Jones, *op. cit.*, I, 156-157.
9. Samuel Butler, "Preface," *op. cit.*, p. ix.
10. Samuel Butler, Letter to Charles Darwin, May 30, 1872 (in Jones, *op. cit.*, I, 157).
11. *Erewhon*, p. 290.
12. May Kendall, *That Very Mab* (London, 1885).
13. *Ibid.*, p. 23.
14. *Ibid.*, p. 40.
15. *Ibid.*, p. 55.
16. *Ibid.*, pp. 88-90.
17. *Ibid.*, pp. 122-123.
18. *Ibid.*, pp. 153-155.
19. *Ibid.*, pp. 157-158.
20. *Ibid.*, p. 159.
21. *Ibid.*, p. 207.
22. Thomas Longueville, *The Life of A Prig* (New York, 1886).
23. *Ibid.*, p. 119.
24. *Ibid.*
25. The full title is: A Full and Free Account of the Wonderful Mission of Earl Lavender, which leaked out one night and one day; with a history of the pursuit of Lord Lavender and Lord Brumm by Mrs. Scamler and Maud Emblem.

26. Hayim Fineman, *John Davidson* (Philadelphia, 1916), p. 48.
27. *Ibid.,* p. 30.
28. *Earl Lavender,* p. 48.
29. *Ibid.,* p. 97.
30. *Ibid.,* p. 139.
31. *Ibid.,* p. 283.

CHAPTER SEVEN

LOSS OF FAITH THROUGH EVOLUTION AND DARWINISM

1. Andrew D. White, *A History of the Warfare of Science with Theology,* I, 71.
2. Edward Maitland, *The Pilgrim and the Shrine* (London, 1871).
3. *Ibid.,* p. 29.
4. Winwood Reade, *The Outcast* (London, 1875).
5. F. Legge, Introduction to *Martyrdom of Man* by Winward Reade (London, 1919), p. viii.
6. W. Reade, Preface to *Martyrdom of Man,* p. 4.
7. *Ibid.*
8. Reade, *The Outcast,* p. 48.
9. *Ibid.* p. 85.
10. James Franklin Fuller, *John Orlebar, Clerk* (London, 1878).
11. *Ibid.,* p. 36.
12. *Ibid.,* p. 288.
13. George Somes Layard, *Mrs. Lynn Linton* (London, 1901), p. 310.
14. Eliza Lynn Linton, *Under Which Lord?* (Leipzig, 1880).
15. *Ibid.,* I, 61.
16. *Ibid.,* I, 133.
17. *Ibid.,* I, 156.
18. *Ibid.,* I, 317.
19. *Ibid.,* II, 236.
20. *Ibid.,* II, 237-239.
21. *Ibid.,* II, 268.
22. Eliza Lynn Linton, *The Autobiography of Christopher Kirkland* (London, 1885).
23. *Ibid.,* III, 79-80.
24. *Ibid.,* III, 186.
25. *Ibid.,* III, 94.
26. George Gissing, *Born in Exile* (London and Edinburgh, 1893), pp. 117-118.
27. George Gissing, *Workers in the Dawn,* ed. by Robert Shafer (Garden City, New York, 1935).

28. Robert Shafer, Introduction to *Workers in the Dawn,* by George Gissing, p. xx, n.
29. Gissing, *Workers in the Dawn,* I, 211.
30. *Ibid.*, I, 216.
31. Richard Dowling, *Under St. Paul's* (London, 1880).
32. *Ibid.*, I, 221.
33. *Ibid.*, I, 250.
34. *Ibid.*, II, 194.
35. *Ibid.*, II, 195.
36. William Hale White, *The Autobiography of Mark Rutherford* (New York, 1916).
37. *Ibid.*, p. 76.
38. Mrs. Humphry Ward, *Robert Elsmere* (New York, 1888).
39. W. F. Monypenny and G. E. Buckle, *The Life of Benjamin Disraeli,* II, 104-105.
40. Ward, *Robert Elsmere,* p. 201.
41. *Ibid.*
42. *Ibid.*, p. 325.
43. W. L. Phelps, "The Novels of Mrs. Humphry Ward," *The Forum* (1909), XLI, 325.
44. Joseph Hocking, *Jabez Easterbrook* (London, 1895).
45. *Ibid.*, p. 13.
46. *Ibid.*, p. 97.
47. *Ibid.*, p. 107.
48. *Ibid.*, p. 263.
49. George Du Maurier, *Trilby* (New York, 1895).
50. *Ibid.*, p. 275.
51. *Ibid.*, p. 283.
52. *Ibid.*, pp. 277-278.
53. *Ibid.*, pp. 285-286.
54. George Moore, *Evelyn Innes* (New York, 1927).
55. *Ibid.*, p. 300.
56. *Ibid.*, p. 354.
57. *Ibid.*, pp. 359-360.
58. *Ibid.*, p. 363.
59. Julia Kavanagh, *John Dorrien* (New York, 1875).
60. *Ibid.*, p. 327.
61. E. C. Grenville-Murray, *That Artful Vicar* (Leipzig, 1879).
62. *Ibid.*, I, 211.
63. *Ibid.*, I, 208.
64. *Ibid.*

65. *Ibid.*, I, 209.
66. *Ibid.*, I, 211.
67. Marie Corelli, *Ardath* (New York, n.d.).
68. Marie Corelli, *A Romance of Two Worlds* (New York, n.d.), p. 237
69. *Ibid.*, p. 331.
70. *Ibid.*, p. 335.
71. Corelli, *Ardath,* p. 20.
72. *Ibid.*, p. 392.
73. *Ibid.*, p. 395.
74. *Ibid.*, p. 435.
75. *Ibid.*, p. 440.
76. Marie Corelli, *The Mighty Atom* (Philadelphia, 1896).
77. *Ibid.*, p. 140.
78. *Ibid.*, p. 141.
79. Eliza Lynn Linton, *op. cit.*, I, 133.
80. Corelli, *Ardath,* p. 436.

CHAPTER EIGHT

COMPROMISE AND CONCILIATION

1. Francis Darwin, ed., *op. cit.*, II, 82.
2. Leonard Huxley, *op. cit.*, II, 401.
3. Charles Kingsley, *His Letters and Memories of His Life,* p. 327.
4. Charles Kingsley, "Natural Theology of the Future" (in *Works,* Vol. XIX), London, 1885, p. 330.
5. Kingsley, *His Letters and Memories,* p. 378.
6. *Ibid.*, p. 337.
7. *Ibid.*, p. 340.
8. Kingsley, "Natural Theology of the Future" (in *Works,* Vol. XIX), p. 332.
9. Kingsley, *Water Babies* (Philadelphia, 1899).
10. *Ibid.*, p. 210.
11. *Ibid.*, p. 66.
12. Sir James C. Lees, *Stronbuy* (Edinburgh, 1883).
13. *Ibid.*, p. 227.
14. *Ibid.*
15. Laurence Oliphant, *Altiora Peto* (Leipzig, 1883).
16. *Ibid.*, II, 23.
17. *Ibid.*, II, 69.
18. *Paul Nugent, Materialist,* p. 223.
19. *Ibid.*, p. 119.

20. *Ibid.*, p. 264.
21. *Ibid.*, p. 273.
22. Ibid.
23. *Ibid.*, p. 339.
24. W. H. Mallock, *Memoirs of Life and Literature,* p. 90.
25. W. H. Mallock, *The New Republic* (London, 1906).
26. *Ibid.*, p. 133.
27. *Ibid.*, p. 112.
28. W. H. Mallock, *The Veil of the Temple* (New York, 1904).
29. *Ibid.*, p. 66.
30. *Ibid.*, p. 236.
31. *Ibid.*, p. 263.
32. Hugh Westbury, *Frederick Hazzelden* (London, 1887).
33. *Ibid.*, p. 64.
34. *Ibid.*
35. *Ibid.*, p. 80.
36. *Ibid.*, p. 81.
37. *Ibid.*, p. 173.
38. Theo. Gift, *Dishonored* (New York, 1890).
39. *Ibid.*, p. 195.
40. *Ibid.*, p. 209.
41. See *ante,* chap. V.
42. George Gissing, *Born in Exile* (London and Edinburgh, 1893).
43. *Ibid.*, pp. 348-349.
44. Mrs. Humphry Ward, *Eleanor* (New York, 1900).
45. *Ibid.*, II, 169.
46. *Ibid.*, II, 424.
47. *Ibid.*, II, 435.
48. Edmund Gosse, *Father and Son* (London, 1930).
49. *Ibid.*, p. 105.
50. *Ibid.*, p. 120.
51. *Ibid.*, p. 110.
52. Edmund Gosse, *Life of Philip Henry Gosse* (London, 1890), pp. 335-336.
53. Mallock, *op. cit.*, p. 441.

CHAPTER NINE

THE ANTHROPOLOGICAL ROMANCE

1. H. G. Wells, *The World Set Free* (London, 1914), p. 2.
2. Henry Curwen, *Zit and Xoe* (New York, 1889).
3. *Ibid.*, p. 111.

4. H. G. Wells, "A Story of the Stone Age," in *Tales of Space and Time* (London, 1906).
5. J. Compton Rickett, *The Quickening of Caliban* (New York, 1893).
6. *Ibid.*, p. 20.
7. *Ibid.*, p. 142.
8. *Ibid.*, pp. 16-17.
9. J. Provand Webster, *The Oracle of Baal* (Philadelphia, 1896).
10. *Ibid.*, p. 88.
11. *Ibid.*, p. 181.
12. *Ibid.*, p. 200.
13. *Ibid.*, p. 303.
14. *Ibid.*, p. 364.
15. *Ibid.*

CHAPTER TEN

THE ROMANCE OF ECCENTRIC EVOLUTION

1. Jules Verne, *A Journey to the Center of the Earth* (London, 1923).
2. Robert Ellis Dudgeon, *Colymbia* (London, 1873).
3. W. G. Bekker, *An Historical and Critical Review of Samuel Butler's Literary Works* (Rotterdam, 1925), p. 90.
4. *Colymbia*, pp. 36-37.
5. *Ibid.*, p. 36.
6. *Ibid.*, p. 237.
7. *Ibid.*, p. 175.
8. William Westall, *A Queer Race* (Philadelphia, n.d.).
9. *Ibid.*, p. 173.
10. H. B. M. Watson, *Marahuna* (London, 1888).
11. *Ibid.*, p. 48.
12. *Ibid.*, p. 192.
13. *Ibid.*, p. 162.
14. *Ibid.*, p. 78.
15. James De Mille, *Strange Manuscript Found in a Copper Cylinder* (London, 1894).
16. *Ibid.*, p. 143.
17. *Ibid.*, p. 72.
18. Lucas Malet, *Colonel Enderby's Wife* (London, 1911).
19. *Ibid.*, p. 4.
20. *Ibid.*, p. 273.
21. Marie Corelli, *Wormwood* (New York, n.d.).
22. *Ibid.*, p. 158.
23. *Ibid.*, p. 222.
24. *Ibid.*, pp. 219-220.

25. Van Wyck Brooks, *The World of H. G. Wells* (New York, 1915), p. 2.
26. H. G. Wells, "The Island of Dr. Moreau," in *Seven Famous Novels* (Garden City, New York, n.d.).
27. H. G. Wells, "Human Evolution—an Artificial Process," *Fortnightly Review,* LXVI (1896), 590-595.
28. *Ibid.,* p. 594.
29. Wells, *op. cit.,* p. 105.
30. *Ibid.,* pp. 155-156.
31. Darwin, *op. cit.,* p. 708.

CHAPTER ELEVEN

EVOLUTION AND THE IDEA OF PROGRESS

1. Thomas H. Huxley, "Government: Anarchy or Regimentation," in *Method and Results—Essays* (New York, 1894), p. 423.
2. Darwin, *The Origin of Species,* pp. 473-474.
3. J. B. Bury, *The Idea of Progress* (London, 1928), p. 336.
4. Q. D. Leavis, *Fiction and the Reading Public* (London, 1932), p. 71.
5. See *ante,* chap. VIII.
6. *Water Babies,* p. 60.
7. *Ibid.,* p. 256.
8. *Ibid.,* p. 96.
9. *Ibid.,* pp. 191-192.
10. *Ibid.,* pp. 184-185.
11. See *ante,* chap. VII.
12. *Autobiography of Christopher Kirkland,* III, 281.
13. Marie Corelli, *The Master Christian* (New York, 1900).
14. *Ibid.,* p. 187.
15. *Ibid.,* p. 174.
16. Marie Corelli, *Temporal Power* (London, 1902).
17. *Ibid.,* p. 304.
18. Robert E. Sencourt, *The Life of George Meredith* (London, 1929), p. 210.
19. Erwin Arthur Robinson, *The Influence of Science upon George Meredith,* Doctoral Dissertation, Ohio State University, 1936, p. 62.
20. George Meredith, Prelude, "The Egoist" (in *The Works of George Meredith,* Vol. XIII), New York, 1910, I, 2-3.
21. *The Egoist,* I, 43.
22. *Ibid.,* I, 45.
23. Edward Clodd, *Grant Allen—A Memoir* (London, 1900), p. 112.
24. Grant Allen, *A Splendid Sin* (New York, 1899).
25. *Ibid.,* p. 98.

26. Marcus Hartog, "Introduction," *Unconscious Memory* by Samuel Butler (New York, 1920), p. xii.
27. Henry F. Jones, *Samuel Butler—A Memoir* (London, 1920), II, 20.
28. Albert Charles Seward, editor, *Darwin and Modern Science* (Cambridge, 1909), p. 88 n.
29. Samuel Butler, *The Way of All Flesh* (New York, 1929).
30. H. F. Jones, *op. cit.*, II, 1.
31. W. T. Young, "George Meredith, Samuel Butler, George Gissing, *The Cambridge History of English Literature* (Cambridge, 1932), XIII, 454.
32. *The Way of All Flesh*, p. 299.
33. H. F. Jones, *op. cit.*, II, 2-3.
34. *The Way of All Flesh,* p. 396.
35. Clara G. Stillman, *Samuel Butler—A Mid-Victorian Modern* (London, 1932), p. 199.
36. *The Way of All Flesh*, p. 389.
37. *Ibid.*, p. 382.
38. *Ibid.*, p. 358.
39. Cyril E. M. Joad, *Samuel Butler* (London, 1924), pp. 161-162.
40. Robert M. Lovett and Helen S. Hughes, *The History of the Novel in England* (Boston, 1932), p. 377.
41. Olive Schreiner, *From Man to Man* (London, 1927).
42. S. C. Cronwright-Schreiner, *The Life of Olive Schreiner* (Boston, 1924), p. 92. See also [S. C. Cronwright-Schreiner] "A Note on the Genesis of the Book," in *From Man to Man*, pp. 19-29.
43. *Ibid.*, p. 215.
44. *Ibid.*, p. 220.
45. *Ibid.*, p. 210.
46. *Ibid.*, p. 435.

CHAPTER TWELVE

EVOLUTION AND THE IDEA OF DEGENERATION

1. Thomas H. Huxley, "Evolution and Ethics" (in *Collected Essays,* Vol. IX), New York, 1905, p. 51.
2. *Ibid.*, pp. 81-82.
3. *Ibid.*, p. 85.
4. Florence Emily Hardy, *The Early Life of Thomas Hardy, 1840-1891* (New York, 1928), p. 198.
5. *Ibid.*, p. 258.
6. *Ibid.*, p. 213.
7. Bruce McCullough, "Introduction" (in *Jude the Obscure,* by Thomas Hardy, Harper's Modern Classics), New York, 1923, p. xxxii.
8. Thomas Hardy, *A Pair of Blue Eyes* (New York, 1929).

9. *Ibid.*, p. 242.
10. Percy White, *Andria* (New York, 1897).
11. *Ibid.*, p. 84.
12. *Ibid.*, p. 274.
13. Frank C. Constable, *The Curse of Intellect* (Boston, 1895).
14. *Ibid.*, p. 98.
15. *Ibid.*, p. 120.
16. *Ibid.*, p. 177.
17. H. G. Wells, *Mr. Blettsworthy on Rampole Island* (Garden City, New York, 1928), p. 204.
18. Lucas Malet, *Colonel Enderby's Wife* (London, 1911).
19. *Ibid.*, p. 150.
20. George Gissing, *New Grub Street* (New York, Modern Library, n.d.).
21. *Ibid.*, p. 488.
22. *Ibid.*, p. 535.

CHAPTER THIRTEEN

THE COMING RACE AND THE IDEA OF PROGRESS

1. Joyce O. Hertzler, *The History of Utopian Thought* (New York, 1926), p. 311.
2. H. G. Wells, "A Modern Utopia" (in *The Works of H. G. Wells,* Atlantic edition, Vol. IX), London, 1925, p. 5.
3. Edward Bulwer Lytton, *The Coming Race* (London and New York, n.d.).
4. Samuel Butler, "Preface," *Erewhon,* pp. vii-viii.
5. Lytton, *The Coming Race,* p. 63.
6. *Ibid.*, p. 71.
7. *Ibid.*, pp. 72-73.
8. Grant Allen, *The British Barbarians* (New York and London, 1895).
9. *Ibid.*, p. 104.
10. *Ibid.*, pp. 206-207.
11. George Du Maurier, *The Martian* (New York, 1897).
12. Dorothy Scarborough, *The Supernatural in Modern English Fiction* (New York, 1917), pp. 252-253.
13. Du Maurier, *Trilby* (New York, 1895), p. 226.
14. Du Maurier, *The Martian,* p. 358.
15. *Ibid.*, p. 375.
16. *Ibid.*, p. 365.
17. *Ibid.*, p. 368.
18. *Ibid.*, p. 405.
19. H. G. Wells, *Experiment in Autobiography* (New York, 1934), p. 461.
20. *Ibid.*, p. 550.

21. H. G. Wells, *op. cit.*
22. Wells, *A Modern Utopia,* p. 7.
23. Charles Darwin, *op. cit.*, p. 706.
24. H. G. Wells, "Men Like Gods" (in *The Works of H. G. Wells.* Vol. XXVIII), London, 1927, p. 106.
25. Wells, *Experiment in Autobiography,* p. 553.

CHAPTER FOURTEEN

THE COMING RACE AND THE IDEA OF DEGENERATION

1. T. H. Huxley, "Evolution and Ethics" (in *Collected Essays,* Vol. IX), p. 85.
2. H. G. Wells, "The Probable Future of Mankind," *Review of Reviews,* LXII (1920), 234.
3. H. G. Wells, *Mr. Blettsworthy on Rampole Island,* p. 205.
4. H. G. Wells, Julian S. Huxley and G. P. Wells, *The Science of Life* (Garden City, New York, 1931), I, 646.
5. Percy Greg, *Across the Zodiac* (London, 1880).
6. *Ibid.*, II, 7.
7. *Ibid.*, I, 282.
8. *Ibid.*, II, 4.
9. W. H. Hudson, *A Crystal Age* (New York, 1916).
10. *Ibid.*, pp. vii-viii.
11. H. G. Wells, "The War of the Worlds," in *Seven Famous Novels* (Garden City, New York, 1934).
12. *Ibid.*, p. 268.
13. H. G. Wells, *Experiment in Autobiography,* p. 433.
14. Wells, *op. cit.*, p. 350.
15. *Ibid.*, p. 351.
16. *Ibid.*, p. 352.
17. H. G. Wells, "The Time Machine" in *Seven Famous Novels* (Garden City, New York, 1935).
18. *Ibid.*, p. 42.
19. *Ibid.*, p. 60.

CONCLUSION

1. H. G. Wells, *Joan and Peter,* p. 27.
2. *The Martian,* p. 369.
3. John Ruskin, "Aratra Pentelice" (in *The Works of John Ruskin,* edited by E. T. Cook and Alexander Wedderburn), London, 1905, XX, 268.

MAITLAND, EDWARD. *The Pilgrim and the Shrine.* London, Chapman and Hall, 1871.

MALET, LUCAS. *See* HARRISON, MARY ST. LEGER KINGSLEY.

BIBLIOGRAPHY

BIBLIOGRAPHY

Novels

ALLEN, GRANT. *The British Barbarians.* New York and London, Putnam, 1895.

"The Child of the Phalanstery" in *The Backslider.* New York, Lewis, Scribner, 1901.

A Splendid Sin. New York, Buckles, 1899.

The Typewriter Girl. New York, Street and Smith, no date.

BEACONSFIELD, EARL OF. "Lothair" in *Novels and Tales* by the Earl of Beaconsfield, Vol. X. London, Longmans, Green, 1900.

"Tancred, or the New Crusade" in *Novels and Tales* by the Earl of Beaconsfield, Vol. IX. London, Longmans, Green, 1900.

BESANT, WALTER, and RICE, JAMES. *Golden Butterfly.* New York, Dodd, Mead, 1888.

BLACKMORE, RICHARD DODDRIDGE. *The Remarkable History of Tommy Upmore.* New York, Harper, 1884.

BOULGER, THEODORA (THEO GIFT). *Dishonoured.* New York, Lovell, 1890.

BRIFFAULT, ROBERT. *Europa.* New York, Scribners, 1936.

BUTLER, SAMUEL. *Erewhon.* Edited by William A. Eddy. New York, Nelson, 1930.

The Way of All Flesh. Modern Readers Series, New York, Macmillan, 1929.

COLLINS, MORTIMER. *Transmigration.* London, Hurst and Blackett, 1874.

COLLINS, WILKIE. "Heart and Science" in *The Works of Wilkie Collins,* Vol. XXV. New York, Collier, no date.

"The Law and the Lady" in *The Works of Wilkie Collins,* Vol. V. New York, Collier, [1875].

CONSTABLE, FRANK CHALLICE. *The Curse of Intellect.* Boston, Roberts, 1895.

CORELLI, MARIE. *See* MACKAY, MARY.

CRAWFURD, OSWALD. *The World We Live In.* London, Chapman and Hall, 1884.

CURWEN, HENRY. *Zit and Xoe.* New York, Harper, 1889.

DAUDET, ALPHONSE. "La Lutte pour la Vie" in *Œuvres Complètes de Alphonse Daudet,* Tome 17. Paris, Houssiaux, 1901, Theatre IV.

DAVIDSON, JOHN. *A Full and True Account of the Wonderful Mission of Earl Lavender.* London, Ward and Downey, 1895.

DE MILLE, JAMES (GILBERT GAUL). *A Strange Manuscript Found in a Copper Cylinder.* London, Chatto and Windus, 1894.

DEMOCRITUS. *Darwin on Trial at the Old Bailey.* London, University Press, 1900.

DOWLING, RICHARD. *Under St. Paul's.* 3 vols. London, Tinsley, 1880.

DOYLE, SIR ARTHUR CONAN. *The Lost World.* New York, Review of Reviews, 1912.

DUDGEON, ROBERT ELLIS. *Colymbia.* London, Trubner, 1873.

DU MAURIER, GEORGE. *The Martian.* New York, Harper, 1897.

Trilby. New York, Harper, 1895.

FRASER-TYTLER, C. C. *See* LIDDELL, CATHERINE.

FULLER, JAMES FRANKLIN. *John Orlebar, Clerk.* London, Smith, Elder, 1878.

GASKELL, ELIZABETH CLEGHORN. *Wives and Daughters.* New York, Harper, 1866.

GAUL, GILBERT. *See* DE MILLE, JAMES.

———, *Gerald Hastings of Barton.* 3 vols. London, Tinsley, 1870.

GIFT, THEO. *See* BOULGER, THEODORA.

GISSING, GEORGE. *Born in Exile.* London and Edinburgh, Black, 1893.

GISSING, GEORGE. *In the Year of Jubilee.* London, Lawrence and Bullen, 1894.

New Grub Street. New York, Modern Library, no date.

The Private Papers of Henry Ryecroft. New York, Modern Library, 1918.

Workers in the Dawn. Edited by Robert Shafer. 2 vols. New York, Doubleday, Doran, 1935.

GOSSE, EDMUND. *Father and Son.* London, Heinemann, 1930.

GREG, PERCY. *Across the Zodiac.* London, Trubner, 1880.

GRENVILLE-MURRAY (EUSTACE CLAIRE). *That Artful Vicar.* 3 vols. Leipzig, Tauchnitz, 1879.

HARRISON, MARY ST. LEGER KINGSLEY (LUCAS MALET). *Colonel Enderby's Wife.* London, Methuen, 1911.

HETHERINGTON, HELEN F., and BURTON, the REVEREND H. DARWIN. *Paul Nugent, Materialist.* New York, Dutton, no date.

HOCKING, JOSEPH. *Jabez Easterbrook.* London, Ward, Lock, 1895.

HUDSON, WILLIAM HENRY. *A Crystal Age.* New York, Dutton, 1916.

HUNT, ROBERT. *Panthea, the Spirit of Nature.* London, Reeve, Benham, and Reeve, 1849.

HUXLEY, ALDOUS. *Brave New World.* Garden City, New York, Doubleday, Doran, 1932.

JENKINS, EDWARD. *Lord Bantam.* New York, Routledge, 1872.

KAVANAGH, JULIA. *John Dorrien.* New York, Appleton, 1875.

KENDALL, MAY E. G. *That Very Mab.* London, Longmans, Green, 1885.

KINGSLEY, CHARLES. *Alton Locke.* New York, Harper, 1861.
Water Babies. Philadelphia, Altemus, 1899.
Westward Ho! Boston, Little, Brown, no date.

LEES, SIR JAMES CAMERON. *Stronbuy.* Edinburgh, Macniven and Wallace, 1883.

LIDDELL, CATHERINE (C. C. FRASER-TYTLER). *Margaret.* New York, Dodd, Mead, no date.

LINTON, ELIZA LYNN. *The Autobiography of Christopher Kirkland.* 3 vols. London, Bentley, 1885.
Under Which Lord? 2 vols. Leipzig, Tauchnitz, 1880.

LONGUEVILLE, THOMAS. *The Life of A Prig.* New York, Holt, 1886.

LYTTON, EDWARD BULWER. *The Coming Race.* London and New York, Routledge, no date.
Kenelm Chillingly. New York, International, no date.
What Will He Do With It? New York, Edward, no date.

MACDONALD, FREDERIKA. *Nathaniel Vaughan, Priest and Man.* New York, Butts, 1874.

MACKAY, MARY (MARIE CORELLI). *Ardath.* New York, Allison, no date.
The Master Christian. New York, Dodd, Mead, 1900.
The Mighty Atom. Philadelphia, Lippincott, 1896.
A Romance of Two Worlds. New York, Hurst, no date.
The Secret Power. Garden City, New York, Doubleday, Page, 1921.
Temporal Power. London, Methuen, 1902.
Thelma. Home Library, New York, Burt, 1906.
Wormwood. New York, Burt, no date.
Ziska. New York, Stokes, 1898.

MALLOCK, WILLIAM HURRELL. *The New Paul and Virginia.* London, Chatto and Windus, 1890.
The New Republic. London, Chatto and Windus, 1906.
The Veil of the Temple. New York, Putnam, 1904.

MEREDITH, GEORGE. "The Egoist" in *The Works of George Meredith,* Memorial edition, Vol. XIII. New York, Scribner's, 1910.

MOORE, GEORGE. *Evelyn Innes.* New York, Appleton, 1927.

OLIPHANT, LAURENCE. *Altiora Peto.* 2 vols. Leipzig, Tauchnitz, 1883.

PEACOCK, THOMAS LOVE. *Gryll Grange.* London, Macmillan, 1896.

Headlong Hall. London, Macmillan, 1896.

Melincourt. London, Macmillan, 1906.

READE, CHARLES. *Put Yourself In His Place.* Library edition. London, Chatto and Windus, 1922.

READE, WINWOOD. *The Outcast.* London, Chatto and Windus, 1875.

RICKETT, J. COMPTON. *The Quickening of Caliban.* New York, Cassell, 1893.

SCHREINER, OLIVE. *From Man to Man.* London, Unwin, 1927.

SMITH, WILLIAM HENRY. *Thorndale; or The Conflict of Opinions.* Edinburgh, Blackwood, 1858.

THACKERAY, WILLIAM MAKEPEACE. *The Newcomes.* Boston, Estes and Lauriat, 1891.

VERNE, JULES. *A Journey to the Centre of the Earth.* London, Milford, 1923.

WALPOLE, F. G. *Lord Floysham.* 2 vols. London, Chapman and Hall, 1886.

WARD, MARY AUGUSTA (MRS. HUMPHRY WARD). *Delia Blanchflower.* New York, Hearst's International, 1914.

Eleanor. 2 vols. New York, Harper, 1900.

Robert Elsmere. New York, Alden, 1888.

The Testing of Diana Mallory. New York, Harper, 1908.

WATSON, H. B. MARRIOTT. *Marahuna.* London, Longmans, Green, 1888.

WEBSTER, J. PROVAND (editor). *The Oracle of Baal.* Philadelphia, Lippincott, 1896.

WELLS, HERBERT GEORGE. "The Empire of the Ants" in *The Works of H. G. Wells,* Atlantic edition, Vol. X. London, Unwin, 1925.

"In the Abyss" in *The Plattner Story and Others.* London, Macmillan, 1920.

"The Island of Dr. Moreau" in *Seven Famous Novels.* New York, Garden City Publishing Company, 1934.

Joan and Peter. New York, Macmillan, 1926.

"Men Like Gods" in *The Works of H. G. Wells,* Atlantic edition, Vol. XXVIII. London, Unwin, 1927.

Mr. Blettsworthy on Rampole Island. Garden City, New York. Doubleday, Doran, 1928.

"A Modern Utopia" in *The Works of H. G. Wells,* Atlantic edition, Vol. IX. London, Unwin, 1925.

"A Story of the Stone Age" in *Tales of Space and Time.* London, Macmillan, 1906.

"The Time Machine" in *Seven Famous Novels.* New York, Garden City Publishing Company, 1934.

"The War of the Worlds" in *Seven Famous Novels.* New York, Garden City Publishing Company, 1934.

The World Set Free. London, Macmillan, 1914.

WESTALL, WILLIAM. *A Queer Race.* Philadelphia, Coates, no date.

WESTBURY, HUGH. *Frederick Hazzleden.* London, Macmillan, 1887.

WHITE, PERCY. *Andria.* New York, Richmond, 1897.

WHITE, WILLIAM HALE. *Autobiography of Mark Rutherford.* New York, Dodd, Mead, 1916.

WHITEING, RICHARD. *The Yellow Van.* New York, Century, 1903.

WRIGHT, SIDNEY FOWLER. *The World Below.* New York and Toronto, Longmans, Green, 1930.

REFERENCE TEXTS

ASQUITH, H. H. *Some Aspects of the Victorian Age* (Romanes lecture, 1918). Oxford, Clarendon, 1918.

AYRES, CLARENCE. *Huxley.* New York, Horton, 1932.

BAILEY, JAMES OSLER. *Scientific Fiction in English, 1817-1914.* A Study in Trends and Forms. Unpublished doctoral dissertation. Chapel Hill, N. C., University of North Carolina, 1934.

BALDWIN, STANLEY E. "Charles Kingsley" in *Cornell Studies in English,* Vol. XXV. Ithaca, N. Y., Cornell University Press, 1934.

BEKKER, W. G. *An Historical and Critical Review of Samuel Butler's Literary Works.* Rotterdam, Van Ditmar, 1925.

BENN, A. W. *The History of English Rationalism in the Nineteenth Century.* London, Longmans, 1906.

BLOOMFIELD, PAUL. *Imaginary Worlds, or The Evolution of Utopia.* London, Hamilton, 1932.

BOSWELL, JAMES. *Boswell's Journal of A Tour to the Hebrides with Samuel Johnson.* Edited by Frederick A. Pottle and Charles H. Bennett. New York, Viking, 1936.

BROOKS, VAN WYCK. *The World of H. G. Wells.* New York, Kennerley, 1915.

BRUNETIERE, FERDINAND. *Le Roman Naturaliste.* Paris, Levy, 1896.

BUCKLE, GEORGE E., and MONEYPENNY, WILLIAM F. *The Life of Benjamin Disraeli, Earl of Beaconsfield.* 2 vols. New York, Macmillan, 1929.

BURY, J. B. *The Idea of Progress.* London, Macmillan, 1928.

BUTLER, SAMUEL. *Evolution Old and New.* London, Hardwicke and Bogue, 1879.

A First Year in Canterbury Settlement. Edited by R. A. Streatfield. New York, Dutton, 1915.

Further Extracts from the Notebooks of Samuel Butler. Edited by A. T. Bartholomew. London, Cape, 1934.

Life and Habit. London, Fifield, 1916.

Luck or Cunning? London, Cape, 1922.

The Notebooks of Samuel Butler. Edited by Henry Festing Jones. New York, Dutton, 1917.

Unconscious Memory. New York, Dutton, 1920.

BYRON, LORD. "Letters and Journals," Vol. V in *The Works of Lord Byron.* Edited by R. E. Prothero. London, Murray, 1922.

CALVERTON, V. F. (editor). *The Making of Man.* New York, Modern Library, 1931.

CAZAMIAN, M. L. *Le Roman et les idées en Angleterre, L'Influence de la Science, 1860-1890.* (Publications de la Faculté des Lettres de l'Université de Strasbourg, Fascicule 15.) Oxford, Milford, 1923.

CHAPMAN, EDWARD M. *English Literature in Account with Religion 1800-1900.* Boston, Houghton Mifflin, 1910.

CHARTERIS, EVAN E. *Life and Letters of Sir Edmund Gosse.* London, Heinemann, 1931.

CLODD, EDWARD. *Grant Allen—A Memoir.* London, Richards, 1900.

COOK, E. T. *The Life of John Ruskin.* London, Allen, 1912.

COURTNEY, JANET E. *Freethinkers of the Nineteenth Century.* London, Chapman and Hall, 1920.

CRONWRIGHT-SCHREINER, S. C. *The Life of Olive Schreiner.* Boston, Little, Brown, 1924.

CRUM, RALPH B. *Scientific Thought in Poetry.* New York, Columbia University Press, 1931.

CUNLIFFE, JOHN W. *English Literature in the Twentieth Century.* New York, Macmillan, 1933.

Leaders of the Victorian Revolution. New York and London, Appleton-Century, 1934.

DARK, SIDNEY. *The Outline of H. G. Wells.* London, Parsons, 1922.

DARWIN, CHARLES. *The Descent of Man and Selection in Relation to Sex.* Reprint of second English edition. Home Library, New York, Burt, no date.

Charles Darwin's Diary of the Voyage of H.M.S. Beagle. Edited by Nora Barlow. Cambridge, University Press, 1933.

Origin of Species. Reprint of sixth London edition. New York, Lovell, Coryell, no date.

DARWIN, ERASMUS. *Phytologia.* London, Johnson, 1800.

Zoonomia, or Laws of Organic Life. 2 vols. Dublin, Dugdale, 1800.

DARWIN, FRANCIS (editor). *Life and Letters of Charles Darwin.* 2 vols. New York, Appleton, 1888.

DE LA MARE, WALTER. *The Eighteen-eighties.* Cambridge, University Press, 1930.

DOWDEN, EDWARD. "The Scientific Movement and Literature" in *Studies in Literature, 1789-1877.* London, Paul, Trench, Trubner, 1906.

DRACHMAN, J. M. *Studies in the Literature of Natural Science.* New York, Macmillan, 1930.

DRAPER, JOHN W. *History of the Intellectual Development of Europe.* New York, Harper, 1876.

ELIOT, GEORGE. "Shadows of the Coming Race" in *Impressions of Theophrastus Such.* New York, Harper, no date.

EXIDEUIL, PIERRE D'. *The Human Pair in the Work of Thomas Hardy.* Translated from the French by Felix W. Crosse. London, Toulmin, 1930.

FINEMAN, HAYIM. *John Davidson.* Doctoral dissertation. University of Pennsylvania, Philadelphia, 1916.

FREEMAN, A. MARTIN. *Thomas Love Peacock.* New York, Kennerley, 1911.

GEIKIE, ARCHIBALD. *The Founders of Geology.* Second edition. London and New York, Macmillan, 1905.

GOSSE, EDMUND. *The Life of Philip Henry Gosse.* London, Kegan, Paul, 1890.

HALDANE, ELIZABETH. *Mrs. Gaskell and Her Friends.* London, Hodder and Stoughton, 1930.

HARDY, FLORENCE EMILY. *The Early Life of Thomas Hardy, 1840-1891.* New York, Macmillan, 1928.

HERTZLER, JOYCE ORAMEL. *The History of Utopian Thought,* New York, Macmillan, 1926.

HUXLEY, LEONARD. *Life and Letters of Sir Joseph Dalton Hooker.* 2 vols. New York, Appleton, 1918.

Life and Letters of Thomas Henry Huxley. 2 vols. New York, Appleton, 1901.

HUXLEY, THOMAS HENRY. *Darwiniana.* New York, Appleton, 1893.

"Evolution and Ethics" in *Collected Essays,* Vol. IX. New York, Appleton, 1905.

"Government: Anarchy or Regimentation" in *Method and Results—Essays.* New York, Appleton, 1894.

Man's Place in Nature and Other Essays. Everyman's Library, New York, Dutton, 1927.

IRVING, WASHINGTON. *History of New York by Diedrich Knickerbocker.* Philadelphia, Carey, Lea, Blanchard, 1836.

JOAD, CYRIL E. M. *Samuel Butler.* London, Parsons, 1924.

JONES, HENRY F. *Samuel Butler—A Memoir.* 2 vols. London, Macmillan, 1920.

KINGSLEY, CHARLES. *Charles Kingsley: His Letters and Memories of His Life.* Edited by his wife. New York, Scribner, Armstrong, 1877.

"Scientific Lectures and Essays" in *Works,* Vol. XIV. London, Macmillan, 1885.

KNIGHT, WILLIAM. *Lord Monboddo and Some of His Contemporaries.* London, Murray, 1900.

KRAUSE, ERNST. *Erasmus Darwin.* Translated from the German by W. S. Dallas. New York, Appleton, 1880.

LAYARD, GEORGE SOMES. *Mrs. Lynn Linton: Her Life, Letters, Opinions.* London, Methuen, 1901.

LEAVIS, Q. D. *Fiction and the Reading Public.* London, Chatto and Windus, 1932.

LOVETT, ROBERT M., and HUGHES, HELEN S. *The History of the Novel in England.* Boston, Houghton Mifflin, 1932.

MCCARTHY, JUSTIN H. *Reminiscences.* New York and London, Harper, 1899.

MCCRACKEN, ANDREW VANCE. *The Theological Reactions of the Victorian Poets to the Natural Sciences and Evolutionism.* Doctoral dissertation, 1932. University of Chicago, Chicago, 1935.

MALLOCK, WILLIAM HURRELL. *Memoirs of Life and Literature.* New York and London, Harper, 1920.

MANNING, HENRY E., Cardinal (editor). *Essays on Religion and Literature.* 3 vols. London, Longmans, Green, 1865-1874.

MELVILLE, LEWIS. *Victorian Novelists.* London, Constable, 1906.

MERZ, JOHN T. *History of European Thought in the Nineteenth Century.* 4 vols. Edinburgh, Blackwood, 1904-1912.

MORGAN, J. V. *Theology at the Dawn of the Twentieth Century.* Boston, Small, Maynard, 1901.

MURRAY, ROBERT H. *Science and Scientists in the Nineteenth Century.* London, Sheldon, 1925.

OSBORN, HENRY FAIRFIELD. *From the Greeks to Darwin.* New York and London, Scribners, 1927.

PEATTIE, DONALD C. *Green Laurels.* New York, Simon and Schuster, 1936.

PLOWMAN, THOMAS F. *In the Days of Victoria.* London, Lane, 1918.

READE, WINWOOD. *Martyrdom of Man.* 21 edition. London, Kegan, Paul, 1919.

RICE-ORLEY, L. *Poetry of the Anti-Jacobin.* Oxford, Blackwell, 1924.

ROBINSON, ERWIN ARTHUR. *The Influence of Science on George Meredith.* Doctoral dissertation. Ohio State University, 1936.

RUSKIN, JOHN. "Aratra Pentelici" in *The Works of John Ruskin.* Edited by E. T. Cooke and Alexander Wedderburn. Vol. XX. London, Allen, 1905.

RUSSELL, BERTRAND. *Religion and Science.* New York, Holt, 1935.

RUSSELL, FRANCES THERESA. *Satire in Victorian Novel.* New York, Macmillan, 1920.

SANDERS, GERALD DEWITT. "Elizabeth Gaskell" in *Cornell Studies in English.* Vol. XIV. New Haven, Yale University Press, 1929.

SCARBOROUGH, DOROTHY. *The Supernatural in Modern English Fiction.* New York, Putnam's, 1917.

SENCOURT, ROBERT ESMONDE. *Life of George Meredith.* London, Chapman and Hall, 1929.

SEWARD, ALBERT CHARLES (editor). *Darwin and Modern Science.* Cambridge, University Press, 1909.

SHAFER, ROBERT. *Christianity and Materialism.* New Haven, Yale University, 1926.

SPENCER, HERBERT. *Study of Sociology.* New York, Appleton, 1874.

STILLMAN, CLARA G. *Samuel Butler, A Mid-Victorian Modern.* London, Secke, 1932.

THOMSON, J. ARTHUR. *The Outline of Science.* 4 vols. New York and London, Putnam, 1922

THORP, MARGARET F. *Charles Kingsley.* Princeton University Press, 1937.

TREVELYAN, JANET P. *The Life of Mrs. Humphry Ward.* New York, Dodd, Mead, 1923.

VAN DOREN, CARL. *The Life of Thomas Peacock.* London, Dent, 1911.

WARD, MARY AUGUSTUS ARNOLD. *A Writer's Recollections.* New York and London, Harper, 1918.

WELLS, HERBERT GEORGE. *Experiment in Autobiography.* New York, Macmillan, 1934.

Mankind in the Making. New York, Scribner's, 1904.

WELLS, H. G., and HUXLEY, JULIAN S., and WELLS, G. P. *The Science of Life.* 2 vols. Garden City, New York, Doubleday, Doran, 1931.

WHITE, ANDREW D. *A History of the Warfare of Science with Theology in Christendom.* 2 vols. New York, Appleton, 1930.

WILSON, REVEREND S. LAW. *The Theology of Modern Literature.* Edinburgh, Clark, 1899.

WINGFIELD-STRATFORD, SIR ESME. *Those Earnest Victorians.* New York, Morrow, 1930.

WILSON, DAVID ALEC and MACARTHUR, DAVID W. *Carlyle in Old Age.* London, Kegan, Paul, 1934.

WOOD, T. M. *George Du Maurier, The Satirist of the Victorians.* London, Chatto and Windus, 1913.

YOUNG, W. T. "George Meredith, Samuel Butler, George Gissing," *Cambridge History of English Literature,* Vol. XIII.

ARTICLES

CLODD, EDWARD. "Dr. Johnson and Lord Monboddo," *Fortnightly Review,* London, CVII, n.s.CI (1917), 849-862.

GOODALE, RALPH A. "Schopenhauer and Pessimism in Nineteenth Century English Literature," *Publications of the Modern Language Association,* XLVII (1932), 241-261.

GRAY, W. FORBES. "A Forerunner of Darwin," *Fortnightly Review,* London, CXXXI, n.s.CXXXV (1929), 112-122.

LINTON, ELIZA LYNN. "Professor Henry Drummond's Discovery," *Fortnightly Review,* London, LXII, n.s.LVI (1894), 448-457.

MACDONALD, W. L. "Samuel Butler and Evolution," *North American Review*, CCXXIII (1926-1927), 626-637.

WELLS, HERBERT GEORGE. *"Human Evolution*—An Artificial process," *Fortnightly Review*, London, LXVI, n.s.LX (1896), 590-595.

"The Probable Future of Mankind," *Review of Reviews*, London, LXII (1920), 231-234.

WILLCOCKS, M. P. "Samuel Butler: Of the Way of All Flesh," *The English Review*, XXXIX (1924), 524-540.

INDEX